Geophysics

Contents

1. INTRODUCTION 3

2. THE GRAVITY METHOD 4
   General Principles 4
   Some Case Studies 5
   Global Applications of the Gravity Method 7

3. THE MAGNETIC METHOD 10
   General Principles 10
   Magnetic Survey Work in Areas of Continental Crust 11
   Magnetic Studies for Global Geological Purposes 12

4. THE SEISMIC METHOD 20
   Seismic Surveying 20
   Earthquake Seismology 27

ANSWERS 33

FURTHER READING 33

Front cover: Seismic profile of the Beryl Oil Field and adjacent areas.

Acknowledgements

The authors are most grateful to Professor D H Griffiths, Dr R F King and Dr P F Barker of the Department of Earth Sciences, University of Birmingham, for their constructive suggestions and help with sources of information following a reading of the initial manuscript. Much helpful advice has been received from Dr D Livesey, Dr F Spode and Mr J Mansfield. We also wish to acknowledge the willingness of Mobil North Sea Ltd and Esso Petroleum Company Ltd to supply materials from their commercial sources.

The following figures have been based on illustrations from the sources indicated:
Fig. 2.4 and 2.5, *Proceedings of the Yorkshire Geological Society, Vol 34*; Fig. 2.11, Fig. 3.13, Fig. 4.16, Fig. 4.17a, Fig. 4.18, Holmes, A. *Principles of Physical Geology*, Nelson; Fig. 3.5, Dunning, F.W. *Geophysical Exploration*, HMSO; Fig. 3.8, Tarling, D.H. & M.P. *Continental Drift*, Bellandson; Fig. 3.14 *Understanding the Earth*, Artemis Press; Fig. 4.5 *Esso Magazine No. 116*, Winter 1980/81; Fig. 4.15, Fig. 4.17b, *Unit 22* Open University S100.

1
Introduction

'The earth scientist with the crystal ball is the geophysicist'.
Prof. I G Gass, Open University

'Geologists seem willing to accept a wider range of speculation from the geophysicists than from their own kind.'
Prof. S E Hollingworth, 1957

In a lecture to sixth formers during the 1970s, a well-known speaker showed, to the satisfaction of his audience, that the future of civilised man depended on the work of the geological scientist. The lecturer was Sir Kingsley Dunham, who was then the Director of the Institute of Geological Sciences and most of his audience were studying geology to GCE Advanced Level: a clear case of 'preaching to the converted'! Nonetheless, the talk was a good reminder of how heavily the modern world has come to rely on resources from the ground. 'Fossil' fuels, metal ores, raw materials for the chemical industry, nuclear fuels, building materials - most of these are ultimately derived from geological sources, and all are becoming increasingly difficult to find. The most accessible reserves have long since been worked out and the bulk of current stocks lie buried beneath hundreds or thousands of metres of other rocks, or even beneath the sea bed. This is even more true of the likely discoveries of future reserves.

In such circumstances, surface geological exploration alone is of limited value and increasing use is made of geophysical techniques in prospecting for buried geological structures and materials of possible economic value. As the name implies, geophysics is concerned with the application of the methods of physics to the solution of geological problems. Many of the techniques rely on principles which are first learned in lower-school physics, although the equipment itself has now reached a very high level of sophistication.

In addition to exploration techniques, geophysics has made a major contribution to our knowledge of the deep interior of the earth. Much of this information has been gathered at static geophysical observatories where equipment is used to measure earthquakes and variations in the earth's magnetic field. The relatively young theory of 'plate tectonics' has been formulated largely in order to explain naturally occurring phenomena which have been recorded with geophysical equipment, either at observatories or with mobile instruments.

Table 1.1 lists the main geophysical exploration techniques in current use. Of these, the first three also have important applications to global geology and form the main theme of this Unit.

Used with care, the geological interpretations of geophysical data are invaluable. It will be seen, however, that it is often possible to make several geological interpretations from each set of data and it must not necessarily be assumed that the first interpretation to 'fit' is the only correct one.

Table 1.1

METHOD	MAIN APPLICATIONS
GRAVITY	Land or sea surveys for structures which may contain hydrocarbons. Delimiting sedimentary basins and igneous bodies. Applications of theory of isostasy.
MAGNETIC	Land, sea or airborne surveys, usually at reconnaissance level, to determine depths of 'basement' below possible hydrocarbon fuel traps. Search for ore minerals. Applications to understanding of crustal structure.
SEISMIC a) reflection profile	Determining likely structures for hydrocarbon fuel traps. Measurements of sediment thickness on sea bed.
b) refraction surveys	Site investigation for civil engineering projects. Shallow and deep crustal studies.
ELECTRICAL e.g. resistivity	Ground-based surveys for: site investigation; superficial deposit studies; studies of water table depth.
ELECTRO-MAGNETIC	Mostly airborne surveys for ore mineral deposits.
RADIOMETRIC	Search for radioactive minerals such as uranium and thorium ores. Reconnaissance surveys from the air, followed by ground work.
BOREHOLE LOGGING - a wide range of measurements from probes lowered into boreholes	Obtaining information of rock-type, stratigraphy and structure during the course of borehole exploration for oil, gas, coal, water etc. and for academic purposes. Particularly valuable when boreholes are not being cored, for reasons of expense or time.

2

The Gravity Method

General Principles

Most people are familiar with the concept of gravity, irrespective of the truth or otherwise of the story of the apple falling onto Sir Isaac Newton's head! We are all aware of the problems of weightlessness suffered by astronauts in space and many students have tried to measure the earth's gravitational field in the school physics laboratory, by swinging a simple pendulum or by falling-weight techniques. Results of around 9.8 m s^{-2} may be obtained by such experiments, with the accuracy of a few per cent. This is quite impressive for a school experiment but is of little value for exploration purposes, where measurements must be accurate to at least one part in 10^7!

The units given above indicate that we actually measure the acceleration due to the gravity of the earth rather than the force acting between the earth and the measuring instrument. It is not therefore strictly correct to refer to the earth's gravitational attraction, although we often do!

Absolute and relative gravity

'Absolute' gravity is measured by a more elaborate version of the schools' falling-weight equipment, but since the apparatus is cumbersome, measurements are normally restricted to widely spaced centres, e.g. at geophysical observatories. For most gravity survey purposes, knowledge of absolute gravity is less important than being able to measure accurately and speedily the minute variations in gravity from place to place. For this purpose, a variety of gravity meters (or gravimeters) has been developed, most of which work on the principle of a very delicate spring balance, where an increase in the gravitational field results in greater extension of the spring. Absolute gravity values may readily be assigned to gravity meter stations by 'tying-in' to observatories where such determinations have already been made. In the case of the British Isles the main base is at Cambridge.

Units

Since surveyors are looking for very minor changes in the acceleration due to gravity, the usual S.I. unit is far too large and the unit used in geophysics is the gravity unit (g.u.). One gravity unit is equal to 10^{-6} m s^{-2}. (Older literature and commercial geophysicists use the milligal. One milligal equals 10^{-5} m s^{-2}.)

The earth's gravity field

What are the factors which influence the earth's gravity field? By far the most important is the overall attraction exerted by the great mass of the earth, but superimposed on this are other influences, of which four are of relevance to the exploration geophysicist.

The effect of the underlying rocks

The aim of most gravity surveys is to determine the disposition of the various rock types lying at shallow depth beneath the area of the survey. Since the gravity meter is affected by the mass underneath it, if the rocks are unusually heavy or light, then the reading will vary accordingly when compared with readings for surrounding rocks. One of the earliest applications for gravity surveys was in the search for salt domes in the Gulf Coast oilfields of the USA. These cylindrical masses often act as oil traps and are of much lower density than the surrounding rocks. There is therefore a mass deficiency and a 'negative anomaly' is recorded, i.e. the gravity values are lower than those measured on the adjacent rocks.

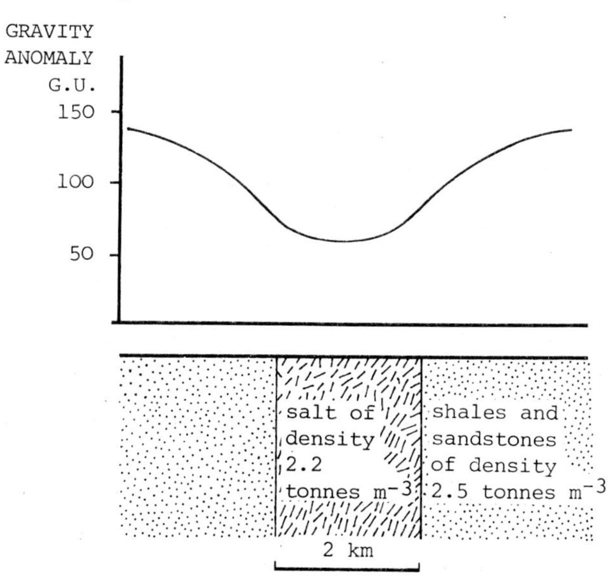

Fig. 2.1 Simplified gravity profile across a salt dome.

Further examples of the applications of gravity surveys will appear later in the Unit. Unfortunately, the effects of the subsurface geology are mostly small compared with the other three influences and corrections have to be made before the survey results are geologically useful.

The latitude effect

It is well known that the earth is not a perfect sphere but more like a rather flattened spheroid. The diameter at the Poles is about 12,714 km in comparison to 12,757 km at the Equator. This means that the surface of the earth at the North and South Poles is nearer to the dense core than is the surface at the Equator. Gravity values at the poles are therefore proportionally higher than those at the Equator. The latitude effect is as great as 15 g.u. per minute of latitude, so very accurate maps are necessary to allow for it in correcting the 'raw' gravity meter readings.

The height effect

The higher one goes above sea level, the further one is from the main mass of the earth and so the gravity readings become less. A modern meter is so sensitive that the difference in gravity between the floor and table height is measurable (3·1 g.u. m^{-1}) so accurately surveyed heights are essential. Unlike the floor-to-table height analogy, in reality, there is rock material between the gravity station and the datum level used for the survey (usually sea level). This has its own attraction which has to be allowed for, using an assumed or measured density for the rock mass. The correction is known as the Bouguer correction and the corrected gravity figures in most common use are referred to as the Bouguer anomaly (Fig. 2.2).

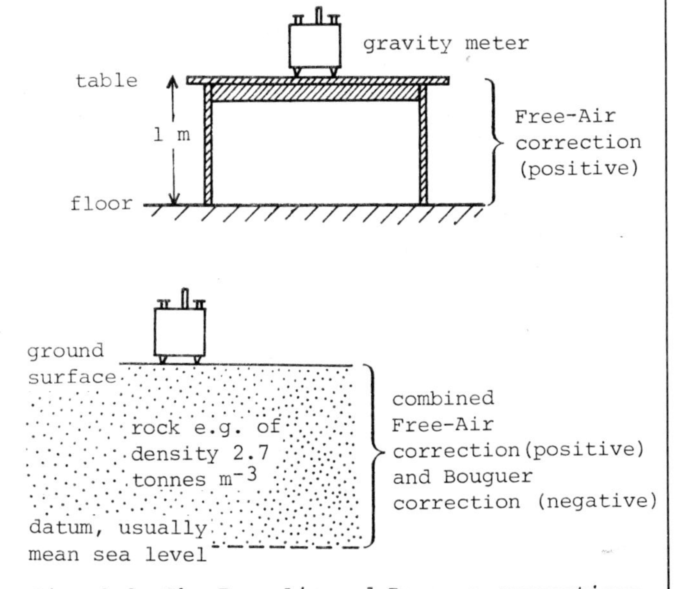

Fig. 2.2 The Free-Air and Bouguer corrections.

The terrain effect

Variations in local relief also exert an influence on the tiny mass in the moving system of the gravity meter. A further correction has therefore to be added. It is calculated using a special grid superimposed on the relief map of the surveyed area.

In very accurate work, the attraction of the sun and moon must also be computed.

Principles of interpretation

The need for data processing described above makes the conduct of a gravity survey sound very laborious. Much of it is rather routine, although computers are now used to speed the calculation procedure. The excitement lies in the interpretation of the Bouguer anomalies!

Before we proceed with some examples, it must be appreciated that the interpretations which usually appear in publications are the best selected from a variety of possibilities. In theory, a large number of underground distributions of rock types might explain a measured Bouguer anomaly. In practice an estimate of what might be happening below ground is drawn up as a 'model'. This is done on the basis of the known surface geology of the area, of measurements of the density of the rocks exposed and on experience of similar geological situations elsewhere. If borehole information is available it is, of course, particularly useful.

The gravity anomaly which would be produced by the 'model' is computed from standard formulae and compared with the measured one. Adjustments are made to the 'model' where there is disagreement and the computation repeated. This may need doing many times until the measured anomaly is matched sufficiently accurately for the purpose of the survey.

Some Case Studies of Interpretations of Gravity Data

An Antarctic glacier

Gravity surveys have been used to estimate the thickness of ice in glaciers for geomorphological purposes, and to add to the accuracy of our knowledge of the total quantity of ice in the world. The example given in Fig. 2.3 is a single gravity profile across the Starbuck glacier in the Antarctic Peninsula (lat. 65° 38' S) beginning and ending on the rock sides. Here a huge density contrast exists between the glacier ice and the rock walls and bed of the valley. Laboratory measurements of the density of glacier ice and of specimens of bedrock (in

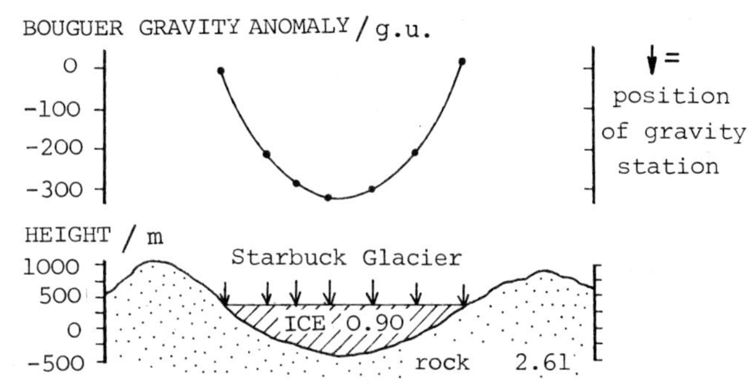

Fig. 2.3 Gravity profile of the Starbuck Glacier. (Vertical and horizontal scales are equal). Densities in tonnes m^{-3}.

this case a granite) gave figures of 0.9 tonnes m^{-3} and 2.61 tonnes m^{-3} respectively, a density contrast of about 1.7 tonnes m^{-3}.

Although there may be density changes within the bedrock below the glacier, they will be small in comparison to the large contrast between the ice and the granite and are unlikely to alter the interpretation very much. The diagram shows a good fit between the measured anomaly shown with a line and that computed for the 'model' shown by points. Evidently there are still some 700 m of ice in the valley, with a further 700 m of towering rock wall above the ice surface.

The Stranraer sedimentary basin

A major application for the gravity method is in delimiting sedimentary basins as an early step in the exploration for oil reserves although the example shown in Fig. 2.4 was surveyed for academic purposes rather than economic ones. It is the Stranraer sedimentary basin in the Southern Uplands of Scotland and it consists largely of Permo-Triassic sandstones and breccias. These lie at right angles to the strike of the dominant Lower Palaeozoic rocks which comprise the rest of the Southern Uplands. A borehole for water had shown that the Permo-Triassic rocks were at least 190 m thick and it was suspected that they were much thicker than this.

The gravity survey revealed a Bouguer anomaly of over 130 g.u. below the background values on the Lower Palaeozoic rocks. Figure 2.5 shows the measured anomaly along the line of section CC', with, beneath it, three possible interpretations of the cross-section of the Permo-Triassic basin. This has been general-

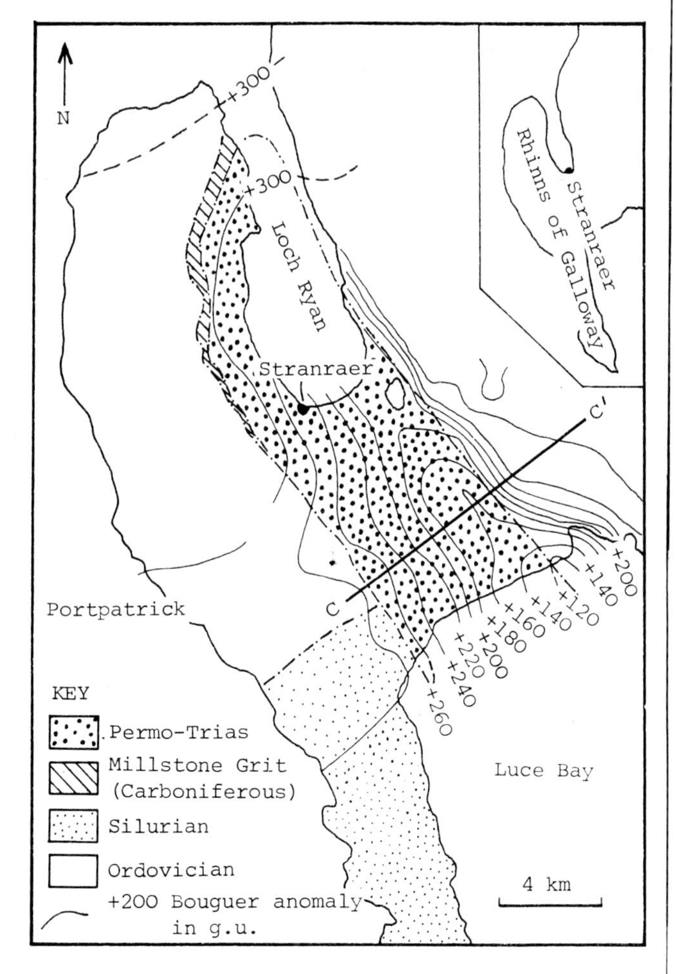

Fig. 2.4 The geology and Bouguer anomalies of the Stranraer district.

ised into a series of steps for ease of computation. A density contrast of 0.4 tonnes m^{-3} (model 2) was considered most probable, from surface measurements, but it is geologi-

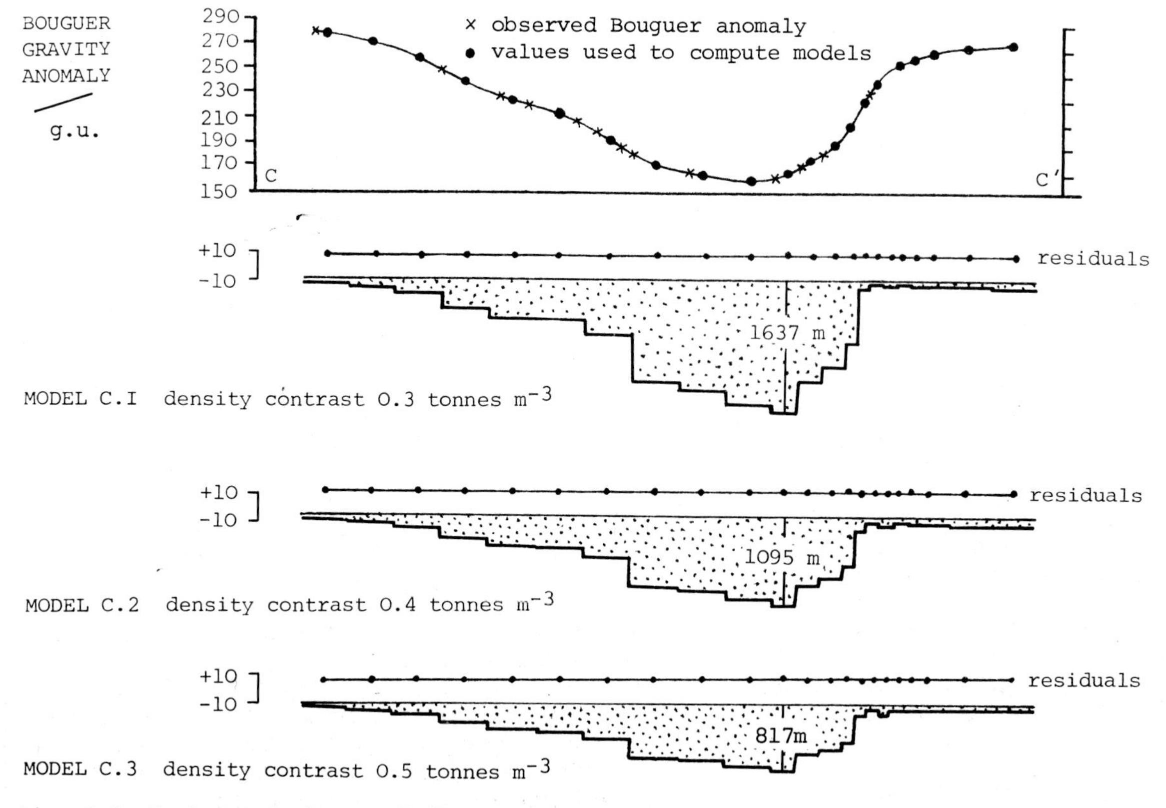

Fig. 2.5 Postulated shapes of the Stranraer sedimentary basin along the line CC' (see Fig. 2.4) for a range of density contrasts.

cally possible that the contrast could be as low as 0.3 tonnes m^{-3} or as high as 0.5 tonnes m^{-3}. The corresponding depths for these density contrasts are shown as models 1 and 3. Even with the uncertainty about the density contrast, the basin proved to be many times deeper than had been thought previously. Furthermore, the gravity gradient was so steep on the eastern side of the basin that it appears almost certain to be due to a normal fault.

Similarly shaped negative gravity anomalies are also common over granites, whose density is usually less than the surrounding country rock, so one must not immediately assume that every low gravity anomaly represents a sedimentary basin.

The Insch Gabbro, Grampian

Before you read on, try to predict the shape of the anomaly which you would expect over a gabbro body, given that gabbro is largely composed of plagioclase (density 2.71 tonnes m^{-3}) and augite (3.3 tonnes m^{-3}) plus possibly olivine (3.7 tonnes m^{3}) whilst granite is mostly orthoclase (2.57 tonnes m^{-3}), plagioclase, quartz (2.65 tonnes m^{3}) and micas (about 2.9 tonnes m^{-3}).

The section in Fig. 2.6 is a north-south gravity profile of the Insch Gabbro and the adjoining Bennachie Granite in Grampian, Scotland. Not suprisingly, the anomaly is higher over the gabbro. A computed 'model' which provides a reasonable fit is given beneath the gravity profile, and shows the gabbro to be a sheet-like intrusion.

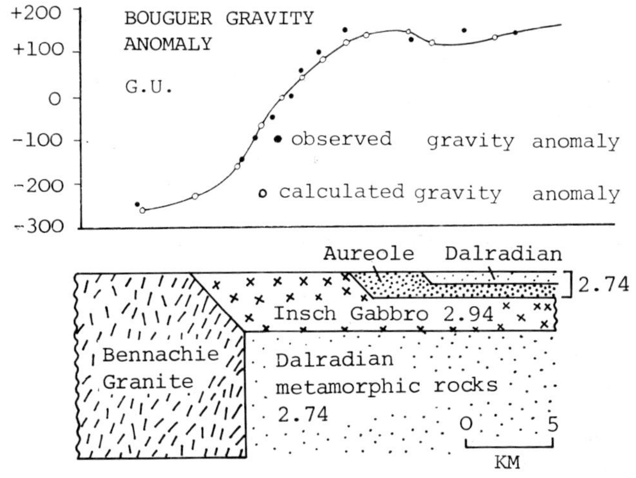

Fig. 2.6 Profile of gravity anomalies across the Insch Gabbro compared with a theoretical profile calculated for the model below. Densities are tonnes m^{-3}.

Global Applications of the Gravity Method

Long before gravity meters had been developed, surveyors had become aware of local irregularities in the earth's gravity field. Indeed, the term 'Bouguer anomaly' is named after a French surveyor, Pierre Bouguer, who worked in the Andes in the eighteenth century. It was not until the middle of the nineteenth century, however, that the implications of changes in the gravity field were fully appreciated and attempts made to match surface measurements with computed variations in the earth's crust. The hypotheses which emerged affect our understanding of major crustal structures today, so a brief account of how they arose is not out of place here.

In the 1840s and 1850s, surveyors were busy mapping accurately the land surface of the Indian sub-continent, mainly to aid the British government in its administration of this huge territory. Two methods were in use, the one complementing the other. Astronomical 'fixes' were painstakingly made at various survey stations, which were then linked by overland measurements using triangulation. In the normal course of events, the results from the two different methods should agree. However, on traverses running north to south, which came within reach of the Himalayan mountain range, considerable errors were noted. For example, the difference in latitude between two stations, Kaliana and Kalianpur (Fig. 2.7) were measured as follows:

by triangulation: 5° 23′ 42".29
by astronomical fixes: 5° 23′ 37".06

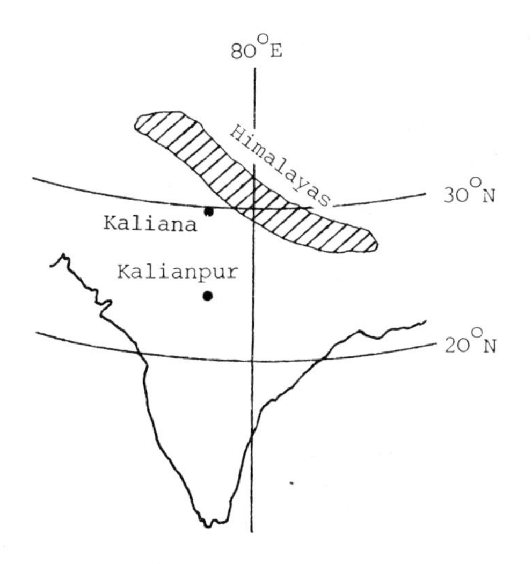

Fig. 2.7 Location of Kaliana and Kalianpur in relation to the Himalayas.

This shows a discrepancy of 5.23 seconds of arc, which does not sound very much, but it represents an error of over 150 metres on the ground. The surveyors knew that their instruments were more reliable than these figures suggested and were not slow to seek an explanation.

The method of position fixing by astronomical means demands a very accurate knowledge of the vertical, whilst triangulation is less dependent on this knowledge. It was therefore suggested that perhaps the Himalayas were exerting a significant sideways pull on the levelling mechanism of the instrument for the Kaliana astro-fix, but that Kalianpur was too distant for it to be so badly affected. This is shown in highly exaggerated form in Fig. 2.8, positions a and b.

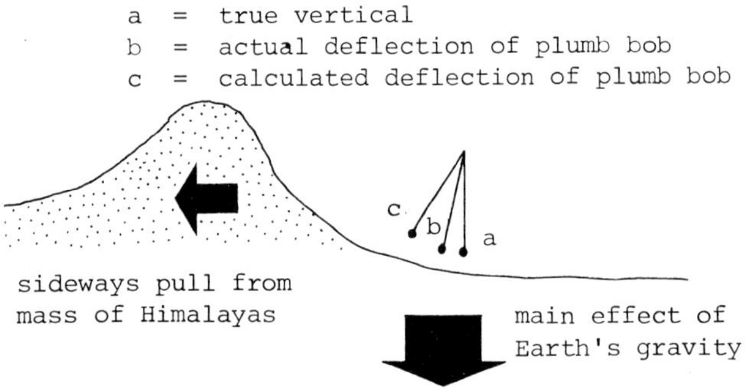

a = true vertical
b = actual deflection of plumb bob
c = calculated deflection of plumb bob

sideways pull from mass of Himalayas

main effect of Earth's gravity

Fig. 2.8 The actual and supposed gravitational effect of the Himalayas on the survey instruments.

Calculations of the probable mass of the Himalayas were made, from their known volume and the density of surface rock samples, but far from matching the observed discrepancy, the calculated deflection produced by the sideways pull of this mass (c in diagram) was three times as great as the observed deflection (b). Why was this?

Up to that time, mountain ranges had been regarded as additional masses 'stuck' on to an otherwise rigid crust of the earth. The density of the crust was assumed to be much the same from place to place. Perhaps these old assumptions were wrong!

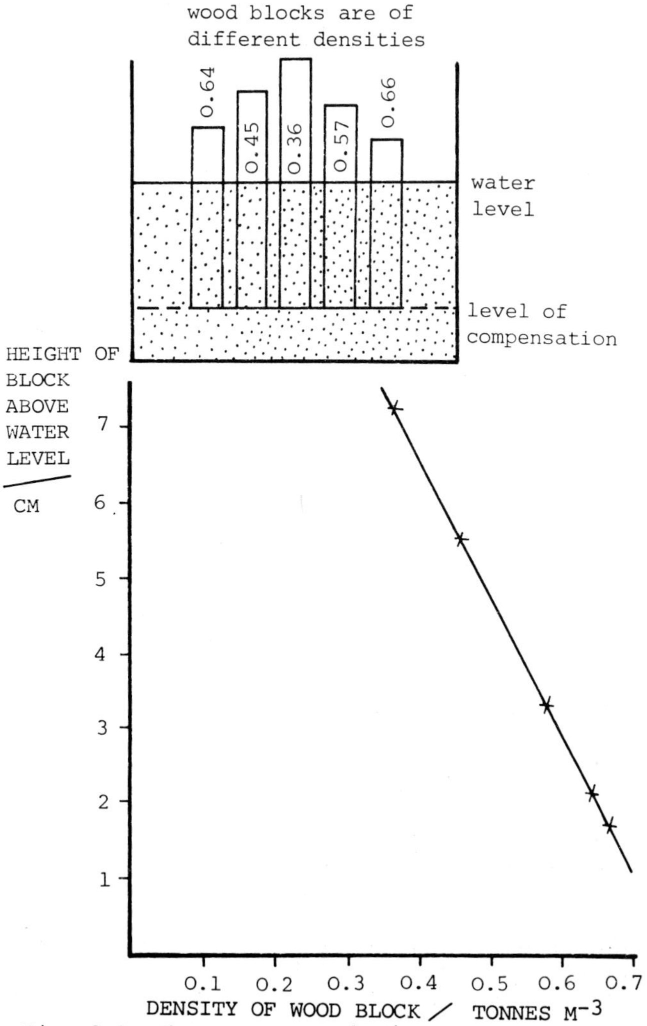

wood blocks are of different densities

0.64 0.45 0.36 0.57 0.66

water level

level of compensation

HEIGHT OF BLOCK ABOVE WATER LEVEL / CM

DENSITY OF WOOD BLOCK / TONNES M^{-3}

Fig. 2.9 The Pratt Hypothesis.

In 1855, two rival hypotheses were published in explanation, each of which proposed that a state of hydrostatic balance existed in the uppermost layers of the earth. Thus J H Pratt suggested that the mountain ranges had risen like fermenting dough - the higher they rose, the less was the density of the rocks composing them. In simple terms, Pratt's hypothesis may be illustrated by floating blocks of wood of varying density in water. The highest blocks are of lowest density and vice-versa. A model to test this idea may be made fairly easily, with the wood blocks free to slide up and down supporting wires to prevent them from toppling over (Fig. 2.9) Note that the bases of the floating blocks lie at an even 'level of compensation', which accorded with current thinking about the crustal layers in the mid-nineteenth century. Beside the diagram is a graph drawn from observations on a home-made model of the height of each block out of the water plotted against its density. *Does the shape of the graph suggest that Pratt's hypothesis is one possible explanation?*

Wood blocks are of the same density (0.62)

water level

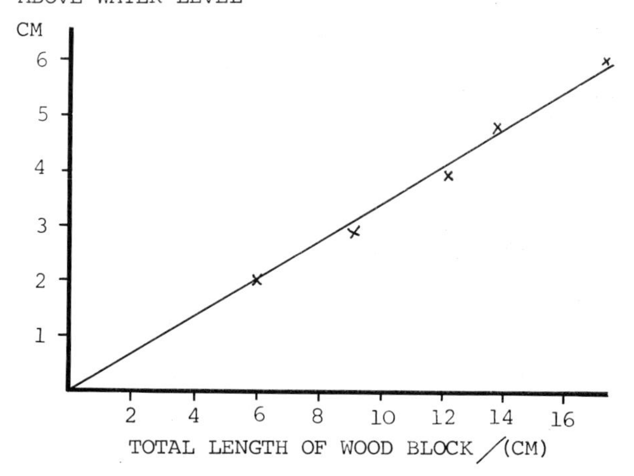

Fig. 2.10 The Airy Hypothesis.

The alternative hypothesis, of G B Airy, proposed that the mountain ranges were behaving rather like icebergs, with a deep 'root' beneath the highest ranges, penetrating into the denser substratum. This idea, too, may be illustrated with wood blocks in water, only here the blocks are all of the same density (Fig. 2.10). Compensation for height is achieved by the block floating more deeply in the water. In this case, the graph shows the height of each block out of the water plotted against its total length. *Does the graph suggest that this hypothesis is also viable?*

Isostasy

Each man had offered an explanation that a state of hydrostatic balance could exist in the earth's surface layers, even though the 'substratum' is not liquid, but rather acts as a plastic substance over long periods of time. Such a state of balance later became known as isostasy (i.e. 'equal stability'), but it was many years before other methods became viable to help test the two hypotheses. Seismic work (described in a later section) has shown that beneath the continents the crust is considerably thicker and of lower density than it is beneath the oceans. To this extent, Airy's hypothesis seems to have the more widespread validity. However, recent seismic work demonstrates that there are some regions, particularly certain mountain ranges, where lateral changes in density offer a better explanation than a deep 'root' of low density material. Thus, Pratt's ideas seem to be vindicated too, albeit on a smaller scale.

Subsequent geophysical work has shown that such a state of isostatic balance does indeed exist throughout much of the world, but there are some exceptional areas where there are huge gravity anomalies, representing considerable mass deficiencies or surpluses and therefore a lack of isostatic balance. Presumably, in such areas, geological processes have affected the crust and upper mantle faster than isostatic adjustments can effect a re-balancing. One area of mass deficiency is North Sweden. This region was one of the last parts of Europe to be deglaciated over the last few thousand years in the wake of the last Ice Age. It would appear that the huge weight of the ice sheet depressed the light rocks of the crust, causing displacement of mantle rocks beneath. The ice then melted very rapidly, in geological terms, and there has not yet been sufficient time for the balance to be re-adjusted by 'flow' within the mantle. There is therefore a mass deficiency beneath North Sweden which is reflected in the negative gravity anomaly and the region is still rising at a measurable rate relative to sea level (Fig. 2.11).

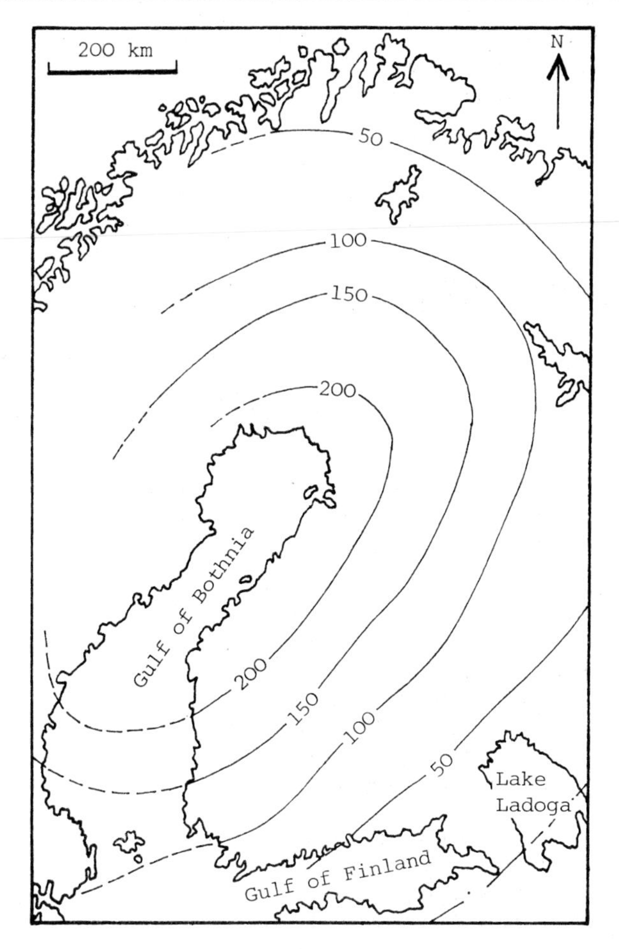

Fig. 2.11 Post-glacial uplift of Fenno-scandia (Finland and Scandinavia). The curves are lines of equal uplift, in metres, from 6800 BC to the present day. Around the northern shores of the Gulf of Bothnia the present rate of uplift is 1 cm per year.

3

The Magnetic Method

General Principles

Most people have used a magnetic compass and know that the magnetised compass needle points north-south. It is also common knowledge that the needle does not point to the geographic north and south poles but rather is aligned towards the magnetic poles, which are several hundred kilometres away from the geographic poles. The exact position of each magnetic pole changes slowly over the years as the magnetic pole 'precesses' around the geographic one. At present in the English Midlands, magnetic north lies some 7° west of true north, decreasing annually. This angle is called the declination (Fig. 3.1).

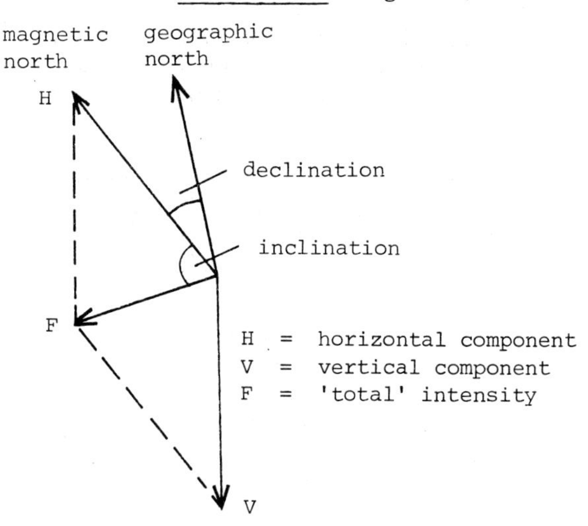

Fig. 3.1 Elements of the earth's magnetic field.

Near the poles themselves, the horizontally-acting component of the earth's field, which has most influence on a compass needle, is very weak. The needle is also being pulled downwards and cannot swing freely, so it becomes very sluggish. In fact, the magnetic pole itself was located by taking measurements with a dip circle, where a magnetised needle is suspended on a horizontal axis. The pole lies beneath the point where the needle reads 90° from the horizontal. On the magnetic equator the dip circle reads zero (Fig. 3.2). The angle from the horizontal of the magnetic dip is known as the inclination (Fig. 3.1).

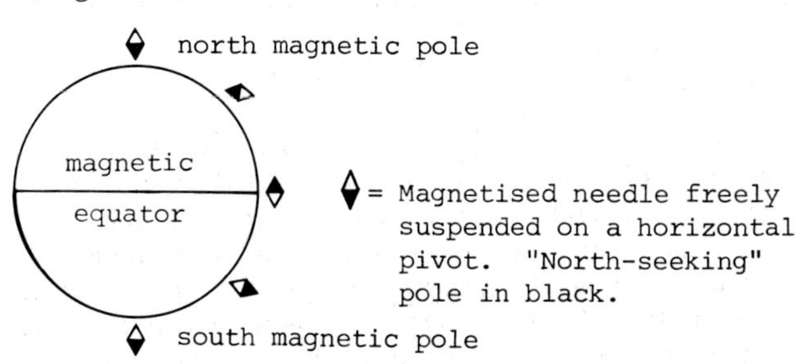

Fig. 3.2 Magnetic latitude.

These well-known instruments are simply responding to the magnetic field of the earth. Experiments are often conducted in the lower school with iron filings or by plotting-compasses around a laboratory bar magnet. Figure 3.3. shows the usual sort of two-dimensional pattern which is produced by iron filings when they are sprinkled on a sheet of card placed over the magnet. The field would be similar if it could be seen in the third dimension too.

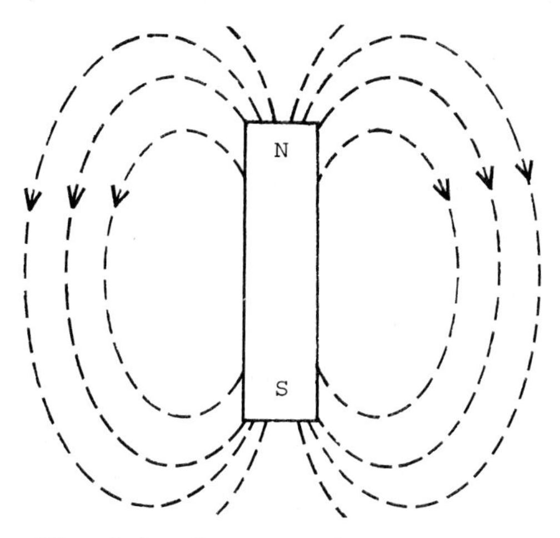

Fig. 3.3 The magnetic field around a bar magnet.

Although no one pretends that the earth has a bar magnet inside it, the magnetic field of the earth is remarkably similar to the one just described. Certainly it is di-polar (i.e. it has a north and south pole) and has apparently been so for most of the earth's history. It probably owes its origin to an internal dynamo effect, produced by thermal convection in the electrically conducting liquid core.

Magnetic field strength

The strength of the earth's magnetic field varies, for three main reasons. First, internal motion in the core produces slow secular changes in field strength, measurable over a period of years. Secondly, there are diurnal changes, brought about by atmospheric disturbances on a day to day basis. Diurnal effects are at their most extreme at times of increased sun-spot activity. Thirdly, and of most interest to exploration geophysicists, variations in near-surface rock types may produce measurable local differences, or anomalies in the magnetic field.

A variety of instruments, known as magnetometers, has been developed to measure such variations in the magnetic field. Earlier instruments mostly depended on delicately pivoted needles, but modern surveyors usually use electronic equipment which is readily adapted for surveys on the ground, in the air or at sea.

Units

The unit used in magnetic work is the nanotesla (nT), based on the S.I. unit, the Tesla. In older surveys the term gamma was used but this is numerically the same as the nT so conversion is easy. The field strength of the earth ranges from about 25,000 nT near the magnetic equator to about 70,000 nT near the magnetic poles.

Some branches of geophysics are concerned with the overall pattern of the earth's magnetic field and maps are available which show the field values and indicate the slow secular changes over the years. Like the earth's gravity, however, there are minor irregularities, or anomalies, superimposed on the main field. Geologically significant anomalies may be up to several thousand nT in amplitude.

Why are there such variations in the magnetic field? There are two main reasons of relevance to the geophysicist.

Induced magnetisation

Many rocks contain iron-bearing minerals, such as magnetite, which are capable of having a magnetisation induced in them by the action of the earth's field today. The ease with which a rock may be magnetised is known as its susceptibility and rocks have characteristic values which may be measured in the laboratory. The susceptibility of most sedimentary rocks is low (but seldom completely negligible). Acid igneous rocks and most metamorphic rocks are about ten times more susceptible, whilst most basic igneous rocks are about one hundred times more susceptible than the sedimentary ones.

Here then, is a potentially useful tool for mapping the sub-surface distribution of crystalline 'basement' rocks, basic igneous bodies like dykes and sills and magnetic ore bodies.

Remanent magnetisation

A second factor, however, must also be considered. This is, the rock may already be 'magnetic' in its own right, resulting from a previous phase in its history, and not connected with its position in the earth's present magnetic field. Such a property is called the remanent magnetisation and it may even be stronger than the induced magnetisation of today. It may also lie in a completely different direction.

The main way in which a rock may acquire such a remanent magnetisation is when it cools from the molten state. When molten, it cannot retain a magnetisation, but once crystallised, it eventually passes a certain critical temperature, known as the Curie Point, below which it can. It then becomes magnetised in the prevailing direction of the earth's field at the time.

Remanent magnetisation may be produced in sedimentary rocks when detrital grains of magnetic minerals settle from suspension. Also, it is common for magnetisation to be produced during the chemical changes which accompany the oxidation of red beds.

Generally, the remanent magnetisation acquired by sedimentary rocks is small in comparison to that in igneous rocks, but there are some circumstances where it is of great interest.

The magnetic anomaly measured at the earth's surface really depends, therefore, on the resultant of two types of magnetisation - induced and remanent (Fig. 3.4).

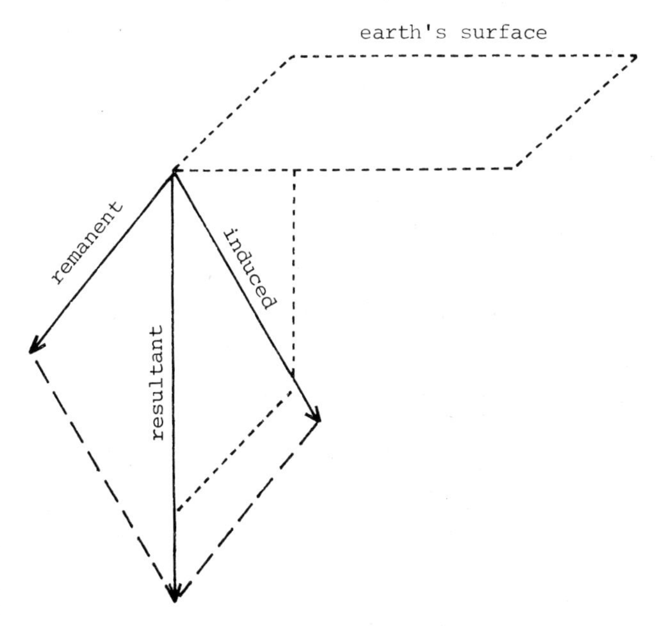

Fig. 3.4 Resultant magnetisation.

Magnetic Survey Work in Areas of Continental Crust

A local survey across an ore body

One of the simplest applications of the magnetic method is in the search for dyke-like bodies of igneous rocks or injections of magnetic ore minerals in otherwise 'non-magnetic' terrain. The example in Fig. 3.5 is from northern Ontario, Canada. A single magnetometer profile across the magnetite veins is shown, together with one possible interpretation of the sub-surface geology. In this case, it is assumed that the anomaly results from the induced magnetisation of the ore body: if its remanent magnetisation is to be considered too, several other interpretations are possible.

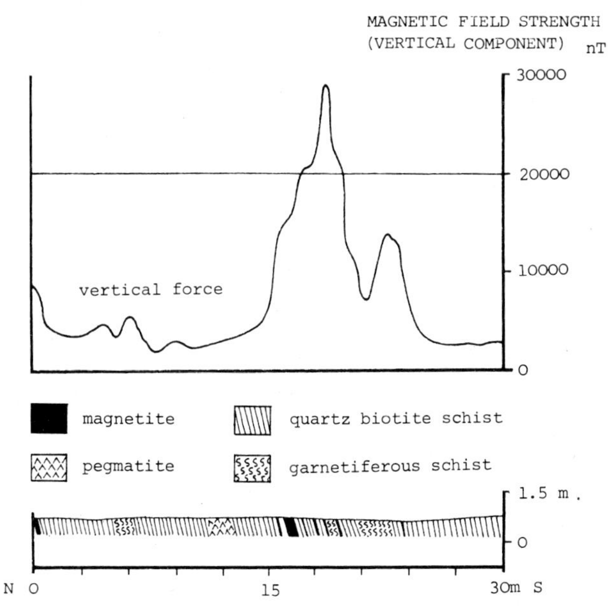

Fig. 3.5 Profile of vertical component of magnetic field across magnetite iron ore layers near Nakina, Northern Ontario, from ground survey.

An aeromagnetic survey

When an oil company begins to explore a new area for hydrocarbon fuels it usually commissions an aeromagnetic survey early in the exploration programme. The method is relatively cheap and fast, and when the aircraft is flying over land it may easily be combined with a programme of aerial photography, for use in structural geological mapping.

Much of the North Sea was covered by airborne magnetometer surveys before further geophysical work was carried out. Figure 3.6 is an extract from an aeromagnetic map and it covers the part of the North Sea where the Beryl Oil Field was subsequently discovered. The location of Block 9/13, which contains the Beryl Field is given in Fig. 4.2.

Figure 3.6 shows that there is a positive anomaly within the block, indicating that magnetic 'basement' rocks are nearer to the surface here than in the immediate surroundings. The steepness of the magnetic gradients suggests that block-faulting is responsible for bringing up the basement rocks and the likely positions of these faults are shown by the double lines in Fig. 3.6.

The axes of the positive anomalies are shown by dotted lines, and depths to magnetic basement have been calculated at points along these axes. Thus, for Block 9/13 the calculated depth is 14,600 feet (4450 metres). Later in the development of this field, holes were drilled, but none of them has yet penetrated the magnetic basement, so its nature in this part of the North Sea is unknown.

The magnetic map itself does not prove the presence of oil, but it does suggest areas where the crystalline 'basement' rocks of the continental crust lie nearer to the surface than usual. These 'basement highs' may relate to structures in the overlying sedimentary rocks which might be favourable for the trapping of oil or natural gas. Again, remanent magnetisation is usually ignored in the commercial companies' interpretations of the magnetic data. The importance of this particular example is pursued in Section 4.

Magnetic Studies for Global Geological Purposes

In the preceding applications of magnetic work, we have mostly considered the induced magnetisation to be the dominant factor in producing a magnetic anomaly. However, since the late 1950s there have been many significant advances in our understanding of the earth, due in no small measure to consideration of the importance of remanent magnetisation. This section will study the measurement of remanence, its application to theories of the history of the continents and lastly its contribution to the magnetic anomalies of the ocean floors. Since remanent magnetisation was produced in the geological past, its study is normally referred to as palaeomagnetism.

Palaeomagnetism and continental drift

Laboratory work is conducted to determine the remanent magnetisation of specimens of 'magnetic' rocks brought back from the field. A portable, petrol-driven rock drill is used to extract small cylinders of rock from the exposure. They are marked with a north arrow and the angle of the specimen from the horizontal is noted, so that it may be re-oriented in the laboratory. The tectonic dip is also noted, so that the effect on the direction of magnetisation of subsequent earth movements may be corrected.

Once in the laboratory, the direction and strength of the remanent magnetisation in the sample is measured by a specially developed magnetometer and an elaborate series of further tests is done to check the reliability of the data. Average values are computed from the results of measurements of several specimens from the same site. After any necessary corrections, the result gives the direction of the earth's field at the time when the rock was magnetised.

Igneous rocks are usually regarded as the most fruitful rock type for such studies, but modern equipment is now delicate enough to be able to cope with the more weakly magnetised sedimentary rocks indicated earlier.

As this kind of research developed in the late 1950s, some rather surprising results began to emerge which have received

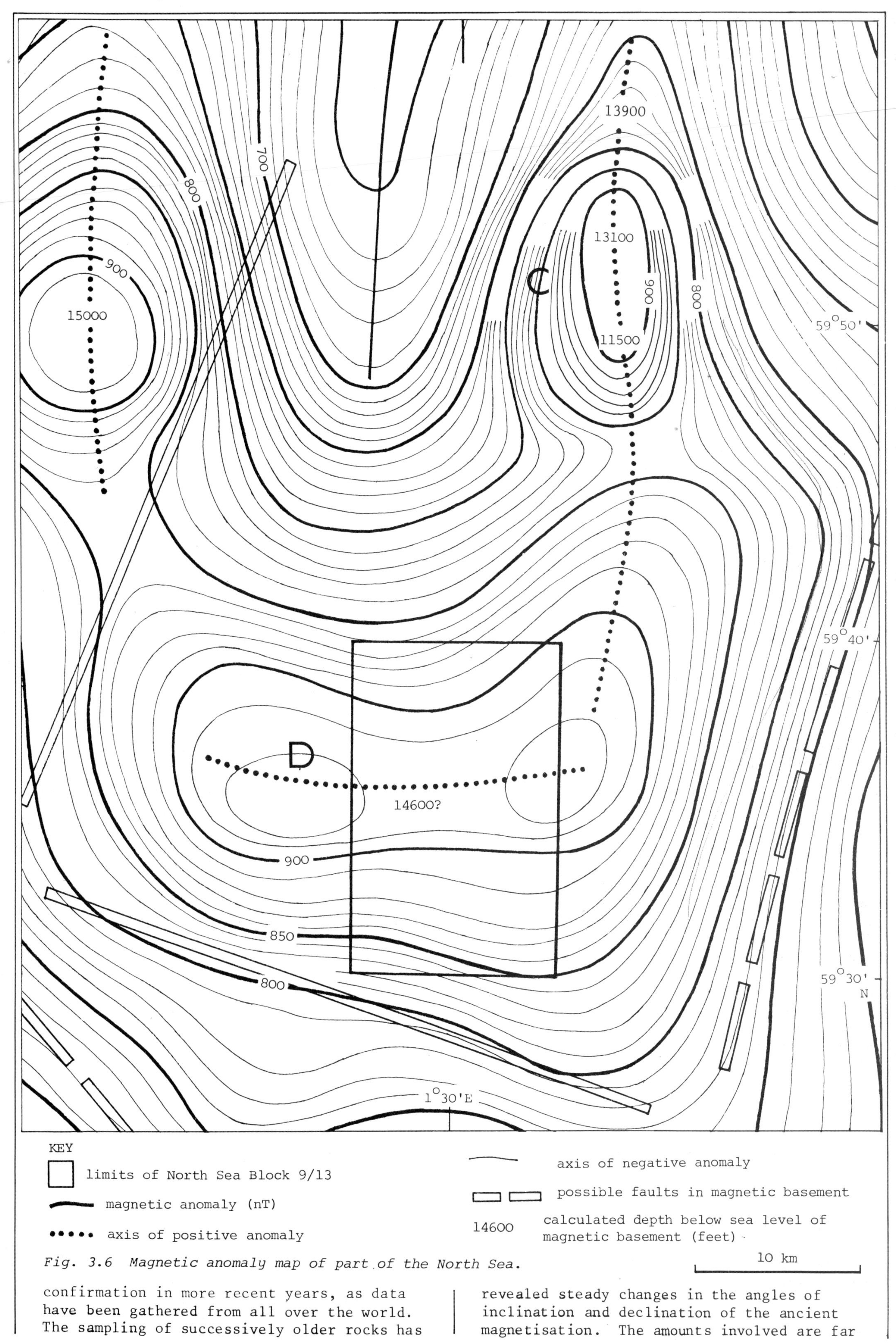

Fig. 3.6 Magnetic anomaly map of part of the North Sea.

confirmation in more recent years, as data have been gathered from all over the world. The sampling of successively older rocks has revealed steady changes in the angles of inclination and declination of the ancient magnetisation. The amounts involved are far

Table 3.1

GEOLOGICAL PERIOD	DATE IN MILLION YEARS	PALAEOMAGNETIC INCLINATION (I)	PALAEO-LATITUDE (L)
Present	0	69°	
Tertiary	50	61°	
Cretaceous	100	57°	
Jurassic	170	50°	
Triassic	220	50°	
Permian	250	23°	
Carboniferous	300	0°	

more than can be explained by the relatively slight precession of the geomagnetic poles around the geographical ones. Indeed, it is generally believed that over long periods of time, the magnetic and geographical poles have had the same average position.

The significance of the changes in direction may be seen from measurements on many British rocks. During the Carboniferous Period, the palaeomagnetic inclination was about 0° from the horizontal, so the British Isles must then have been on the Equator. Working upwards through the succession, the value steadily increased until today, the magnetic dip is about 69°.

Assuming that the earth's field has been dipolar, at least since the Carboniferous, it is possible to plot the changing position of Britain in relation to the Poles. Table 3.1 gives some generalised palaeomagnetic inclinations for Britain at various times since the Carboniferous.

Fortunately there is a simple relationship between the angle of inclination of the remanent magnetisation and the ancient, or palaeo-latitude, expressed in the equation:

Tan I = 2 Tan L, where
I = angle of inclination
L = latitude (Fig. 3.7)

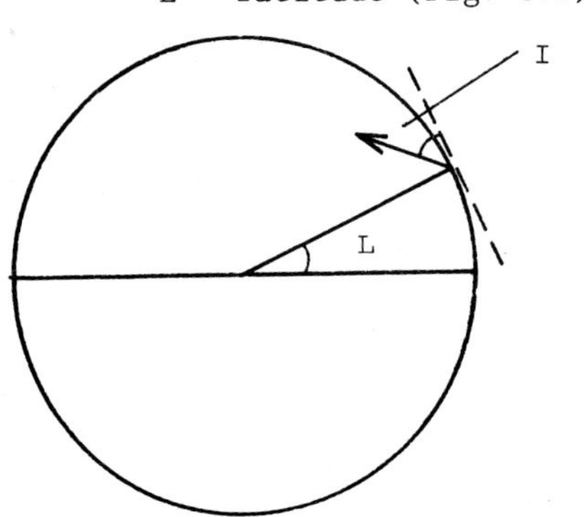

Fig. 3.7 Relationship between latitude (L) and magnetic inclination, or dip (I).

Using this equation, calculate the changing palaeolatitude of Britain and plot the positions on a globe or on an Atlas projection. Note that it is not possible to find the palaeolongitude by this method.

In carrying out this plot, you may have satisfied yourself that you have 'proved' that a portion of the earth's crust may move about on the earth's surface in relation to the poles. On a wider scale, this has become known as the 'Continental Drift Hypothesis'. Have you proved this, however? What if Britain had remained stationary and it was the earth's magnetic axis which had been moving? In the 1950s, the hypothesis of continental drift was not favoured by European and American geologists and the alternative idea, known as the 'Polar Wandering Hypothesis', was given strong consideration. Indeed, you could equally well redraw your map, showing Great Britain at the same position on the globe and a series of points to represent the successive positions of the poles. Joining these points would produce an 'apparent polar wandering curve'.

Later, however, it was shown that the magnetic poles have probably never 'wandered' more than a few hundred miles from the geographic poles and that the latter are unlikely to have moved to anything like the extent required by the polar wandering hypothesis.

As palaeomagnetic results began to be derived from other continents too, the continental drift explanation seemed to be the more plausible. Workers were each plotting their apparent 'polar wandering curves' for their own particular continent, but these frequently disagreed, and only converged onto a single North or South Pole for the present day! Were there, then, dozens of different poles in the past? In spite of the sceptics the simpler solution seemed to be to 'slide' continents about the globe until their 'apparent polar wandering curves' coincided for large periods of geological time (Fig. 3.8). The resultant pattern was remarkably similar to the various attempts to fit the continents together on the basis of the edges of the continental structure at 1000 m below sea level. Further checks were made to see how well the continental reconstructions on the basis of palaeomagnetic latitudes tallied with the ancient climates, distribution of fossils and structural connections known already from geological work. Again the match was remarkably close. Geologists who had worked in the Southern Hemisphere, believing all along in continental drift, now had every reason to be smug, whilst their colleagues in the Northern Hemisphere had the grace to metaphorically eat their earlier words and agree that if everything fitted in so well, then continental drift must, after all, be a 'good thing'!

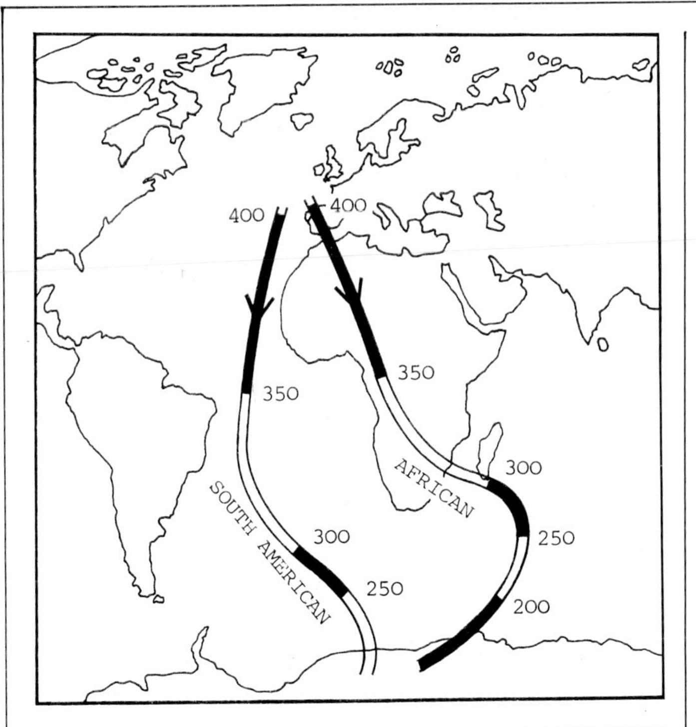

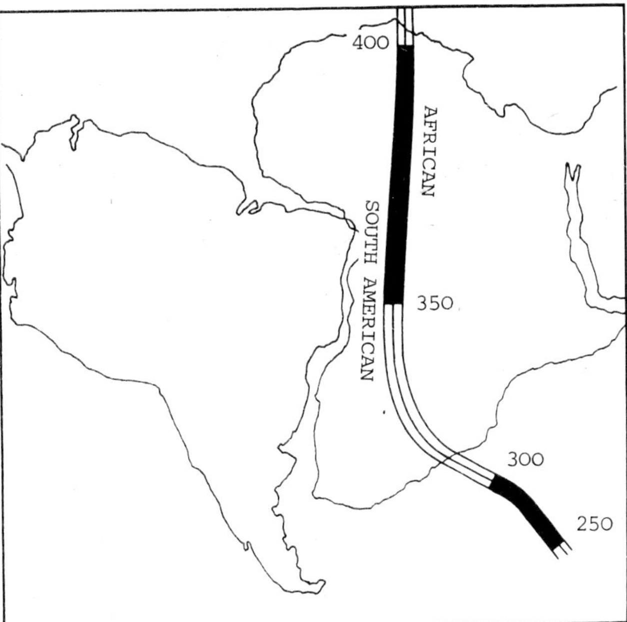

Fig. 3.8 the fit of Africa and South America to reconcile their apparent polar wandering curves: a) The apparent polar wandering curves for the two continents in their present position (dates in millions of years) b) Moving the continents to match their apparent polar wandering curves produces a fit identical to that required by geological observations for at least 150 million years.

Reversals of the earth's magnetic field

We have become accustomed to the north-seeking end of a compass needle always pointing to the North Pole and we would be somewhat surprised if it suddenly pointed South! As the number of palaeomagnetic measurements grew, however, it became clear that such a reversal of polarity had happened, not just once in the past, but many times.

In rocks such as lavas it is also possible to date the different periods of normal or reversed magnetisation by radiometric means (see the Unit *Geochronology*). At first, lavas were collected for this purpose from successive flows in areas such as Iceland and a chronology worked out for the last $4\frac{1}{2}$ million years or so. Beyond that, the radiometric method was not accurate enough. However, once techniques for coring at sea were developed it became possible to collect fossil-bearing sediments immediately overlying the lavas. This has recently enabled the production of the time scale shown in Fig. 3.9, which can accurately be extended back to the Upper Cretaceous, some 70 million years ago.

Many other periods of reversed polarity have also been recognised throughout older parts of the geological time scale.

From the geological record, reversals appear to happen suddenly, but it is probable that the magnetic field dies away to nothing and then builds up again in the opposite direction over a time span of several thousand years.

At first, reversals were regarded as a scientific curiosity, but they later became of great importance in aiding our understanding of the geology of the ocean floors, as the next section shows.

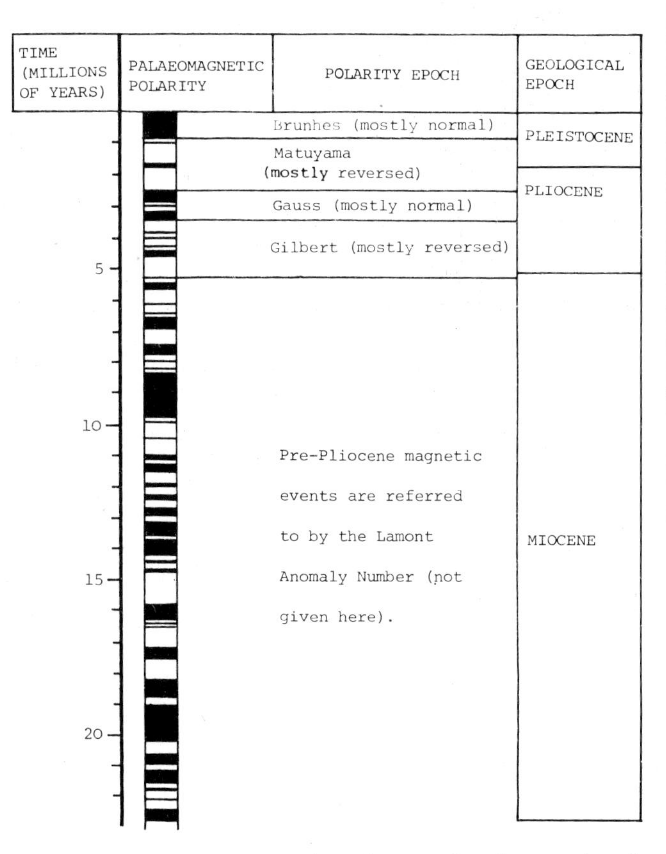

Fig. 3.9 The magnetic time scale for part of the Tertiary and Quaternary. Shaded blocks represent normal polarity; open blocks represent reversed polarity

Magnetic survey work in the ocean basins

During the late 1950s and early 1960s a considerable number of research ships were busy surveying the ocean floors with echo-sounders to determine their relief features. At the same time, each ship could also tow the detector unit of a magnetometer, originally developed as a rather primitive means of submarine spotting! Detailed maps of the type shown in Fig. 3.10 were produced. Rather than showing magnetic contours, as most previous aeromagnetic maps had done, the map depicts positive magnetic anomalies in black and negative ones in white.

The magnetic anomalies revealed a far more complex pattern than many geologists had expected. However, apart from picking out obvious fracture zones in the rocks below the sea bed, no really satisfactory interpretation could be made at first. No one could be sure how much of the anomaly was due to susceptibility contrasts and how much was the result of changes in remanent magnetisation. The rocks of the deep ocean floor were then quite inaccessible, so oriented specimens could not be brought back for laboratory measurements of remanence.

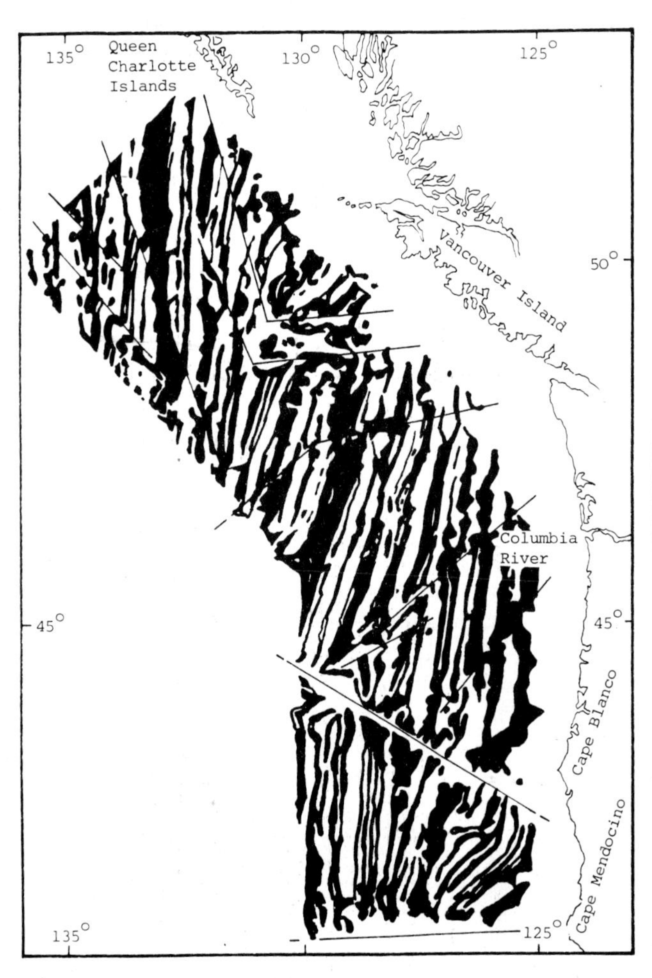

Fig. 3.10 Magnetic anomaly patterns in the north-east Pacific off the Canadian and United States coasts. Straight lines indicate faults displacing the anomaly pattern.

The Vine and Matthews hypothesis

In 1963, two British geophysicists, F J Vine and D H Matthews published a very short paper suggesting a clue to the problem. They pointed out that if the rocks producing the anomalies were of igneous origin, the remanent magnetisation could well be of greater importance than any susceptibility contrast. If the sea bed were underlain by adjacent intrusions or lava flows produced at intervals over several million years, it was quite probable that some would have been formed at times when the earth's magnetic field was normal, and others when it was reversed. Igneous bodies of normal polarity would 'reinforce' the modern earth's field, producing a positive anomaly; bodies of reversed polarity would 'subtract' from it and give rise to a negative anomaly (Fig. 3.11).

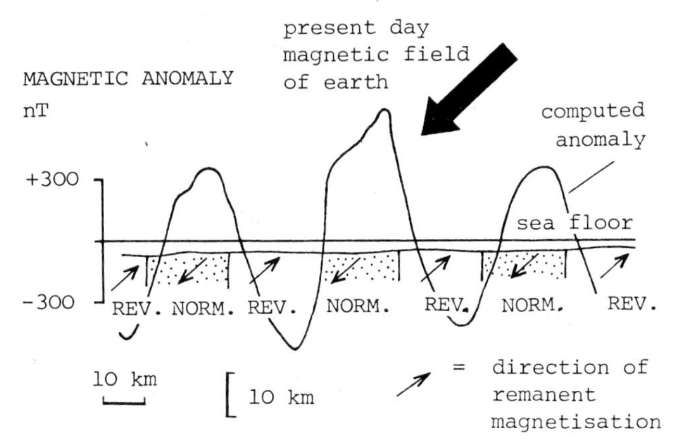

Fig. 3.11 Computed magnetic anomalies across a series of reversed and normally magnetised blocks of oceanic crust.

Supporting evidence for their ideas came from an abortive attempt to drill a hole right through the earth's crust into the mantle, which took place at about this time. Drill cores were recovered which showed reversally magnetised basalts lying at shallow depth below the sea floor.

Sea-floor spreading

The previous year, an American geologist, H H Hess, had attempted to explain the origin of the oceanic ridges, such as the Mid-Atlantic Ridge. He had suggested that beneath the ridges the rocks of the upper mantle were being partly melted, releasing basaltic magmas which were then injected into the crust above or poured out onto the sea bed as lavas.

Once Hess' idea was combined with Vine and Matthews' suggestion of the source of the anomalies, a hypothesis was born which was capable of being tested. Rather than a haphazard arrangement of positive and negative anomalies, it was now realised that they should be symmetrical about the ocean ridges. As each successive injection of magma occurred, it would displace earlier rocks symmetrically to each side. From time to time the earth's field would change polarity and some of the rocks would be reversally magnetised, resulting in alternate positive and negative anomalies, again centred symmetrically about the axis of the ridge. Figure 3.12 shows in diagrammatic form how this might happen. Detailed surveys of portions of oceanic ridges were then

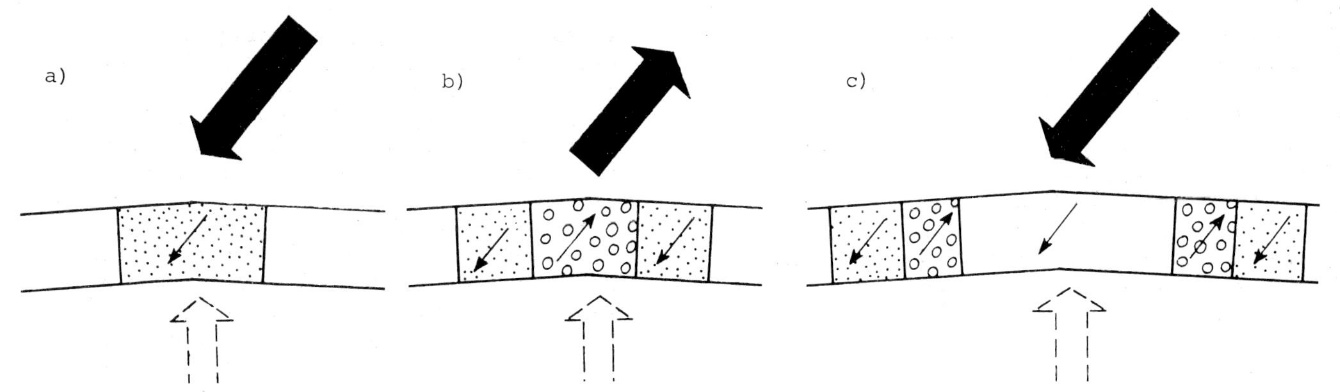

Fig. 3.12 a) First intrusion at ridge axis at time of normal polarity. Remanent magnetisation acquired in same direction as present field. b) Second intrusion at ridge axis at time of reversed polarity. First intrusion pushed apart. New material reversally magnetised. c) Modern intrusion at ridge axis under normal polarity. Earlier intrusions pushed apart. New material normally magnetised.

carried out and the predicted symmetry in the magnetic anomalies soon became apparent. A particularly clear example is found over the Reykjanes ridge, south of Iceland (Fig. 3.13). When the anomalies are plotted using the now conventional black for positive ones and white for negative, they give to the magnetic maps of the ocean floors the appearance of a zebra crossing and the magnetic anomaly belts have become known as 'sea-floor stripes'. The hypothesis that the oceanic crust is generated at the ocean ridges and pushed along like a conveyor belt is known as the 'sea-floor spreading hypothesis'.

Soon after the initial ideas were formulated, it became possible to time the rate of sea-floor spreading. It was noticed that the width of each anomaly was proportional to the length of time determined for each

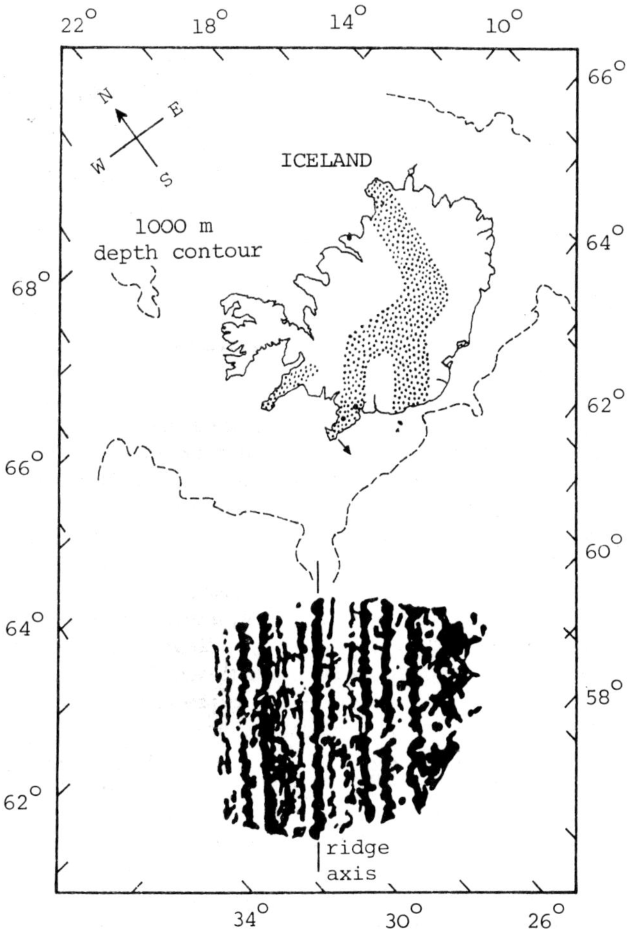

Fig. 3.13 Magnetic anomalies over the Reykjanes Ridge south west of Iceland, showing 'magnetic stripes' (positive anomalies black) and their bilateral symmetry.

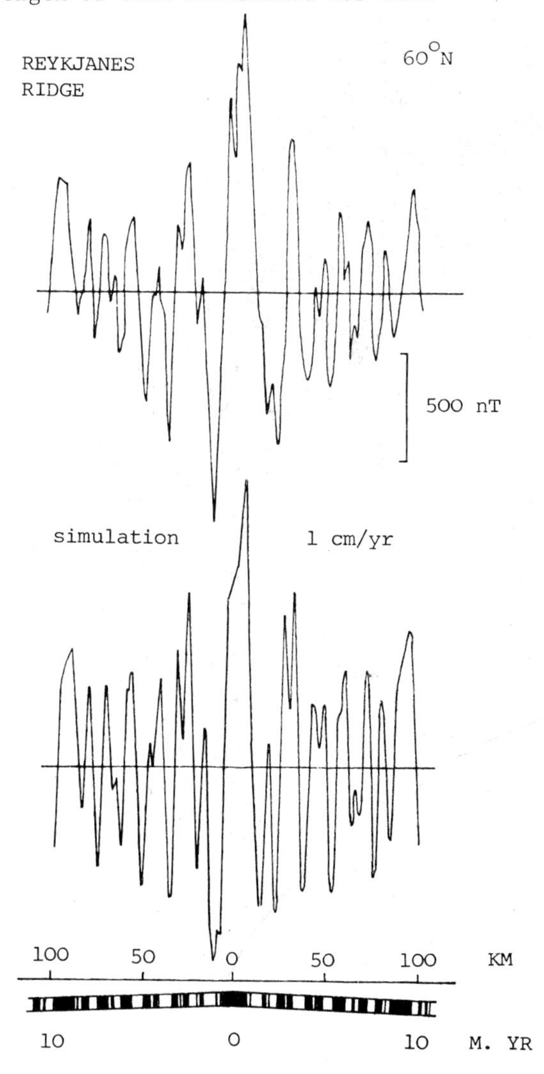

Fig. 3.14 An observed aeromagnetic profile across the Reykjanes Ridge, compared with a computed profile assuming the reversal time scale shown below the profile.

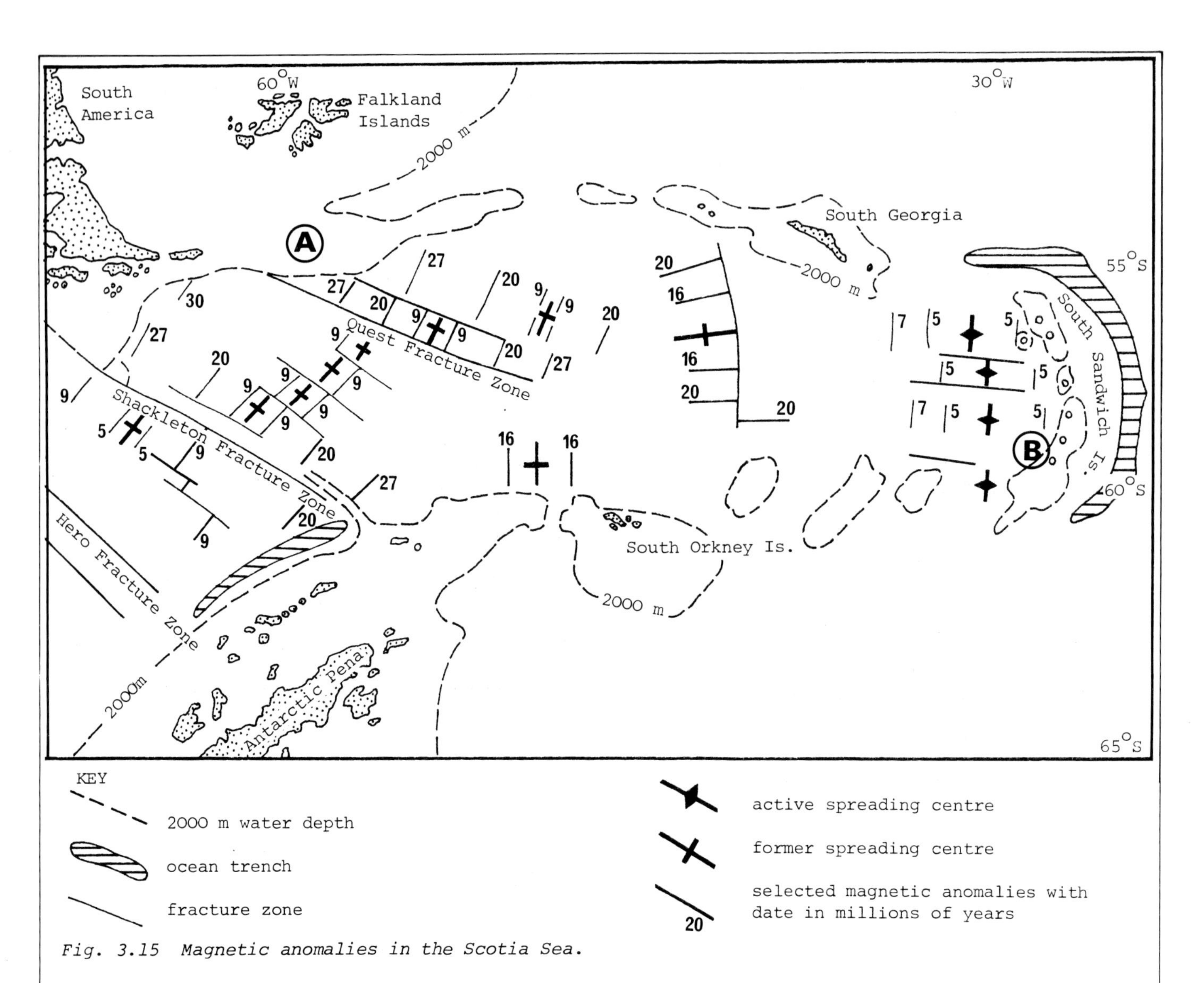

Fig. 3.15 Magnetic anomalies in the Scotia Sea.

period of normal or reversed polarity in Iceland and elsewhere (Fig. 3.9).

Using this time scale, theoretical anomalies were calculated and compared with the anomalies actually measured across the ridges (Fig. 3.14). In most cases the match between calculated and measured anomalies was very good and seemed to point to spreading rates which have remained remarkably constant over long periods of time for each of the major oceanic ridges. Thus, in the North Atlantic, the spreading rate has been a consistent 2 cm per year (i.e. 1 cm per year for each side of the ridge) for a period of several millions of years. In the East Pacific Rise, the rate is higher, being about 10 cm per year.

Such was the rapid acceptance of the sea-floor spreading hypothesis that it was followed by a flurry of activity among sea-going geologists. Existing records which had only been partly interpreted were re-examined and fresh surveys were conducted in critical areas of the oceans. It is now appreciated that each of the 'active' oceanic ridges (i.e. ridges where there is also volcanic and earthquake activity) exhibit the characteristic symmetrical magnetic anomalies and the hypothesis is firmly established in geological thinking.

The Scotia Sea

The Scotia Sea provides just one example of the way in which the sea floor spreading hypothesis has affected our interpretation of geophysical data.

The Scotia Sea lies within the Scotia Arc - a chain of islands and submarine ridges which appear to continue a structural link between South America and Antarctica (Fig. 3.15). Prior to 1963, four summer seasons had been devoted to a magnetometer survey of the Sea. The only analysis which could usefully be attempted was statistically to define magnetically 'quiet' areas, in contrast to more 'active' ones. Thus Area A (Fig 3.15) showed few anomalies and was equated with a great thickness of sedimentary rocks forming the sea bed. By contrast, the sea floor near the South Sandwich Islands (Area B, Fig. 3.15) contains many sharp anomalies, and was regarded as volcanic in origin. Little could be done to relate together the various aspects of the Arc, nor to explain its geological history.

After the publication of the sea floor spreading hypothesis, subsequent expeditions to the Scotia Sea have looked for, and found, symmetrical anomalies similar to those about the active ridges of the major oceans. A typical magnetic profile is shown in Fig. 3.16a. The main map (Fig. 3.15) shows the positions of some more selected anomalies, with the dates in millions of years, derived from the reversal time scale in Fig. 3.9. The most striking feature is an active spreading centre in the east Scotia Sea, but there are other older and now inactive spreading centres in the central and western parts of the area. In addition, several major fracture zones have been discovered from the magnetic anomalies.

The full interpretation of the data is still not complete, but much has already been learned about the geological history of the region and the timing of the break-up of Antarctica, South America and Australia from the super-continent of Gondwanaland of which they had once been part.

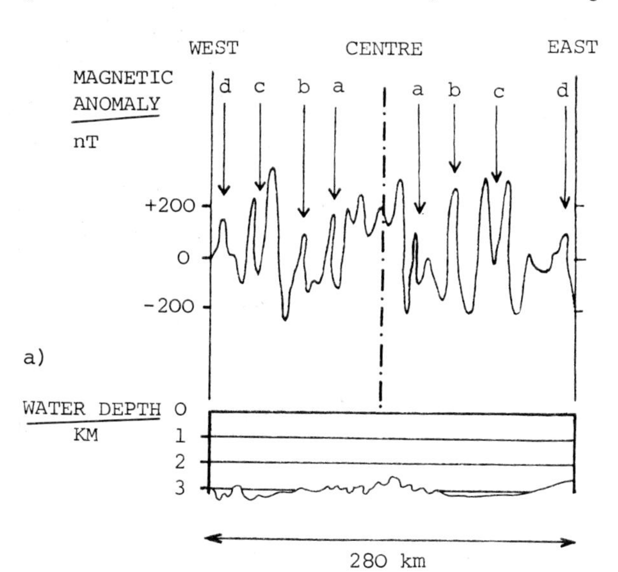

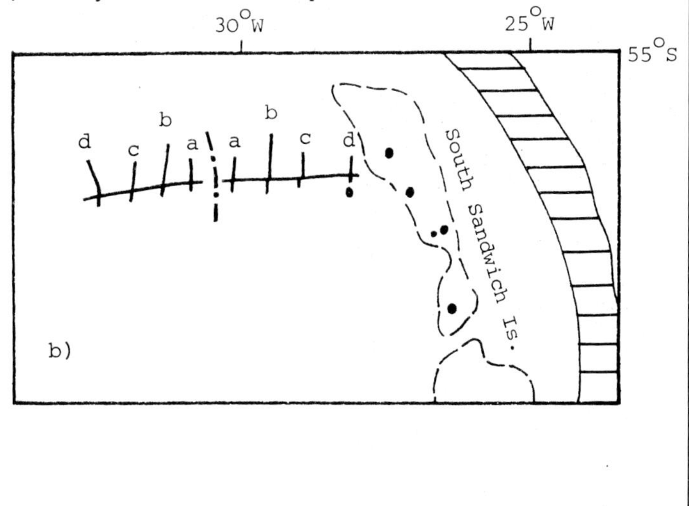

Fig. 3.16 a) Magnetic and bathymetric profile in the eastern Scotia Sea. Anomalies which are symmetrical about the central ridge are lettered. b) Location of profile and major anomalies shown in a).

4

The Seismic Method

For a variety of economic purposes, aeromagnetic and gravity surveys are used early in the exploration of a region, usually because of the speed of the survey and the relatively low cost. Unlikely areas may thus be eliminated and the search concentrated on potentially fruitful underground structures. It is then that the more expensive seismic techniques come into use. Here, explosives or other man-made sources release energy into the ground and sensitive detectors record the return to the surface of reflected or refracted waves.

Much of our knowledge of the interior structure of the earth comes also from studies of naturally occurring earthquakes, but first we shall examine some of the methods of seismic exploration, using controlled sources of energy.

Seismic Surveying

Reflection techniques

An everyday example of a sort of 'seismic' reflection technique is the ship's echo sounder, where a shock wave is transmitted several times a second from a transducer mounted in the ship's hull. The resulting shock wave travels out in all directions through the water, but part of the energy is reflected back to the ship from the sea bed and the time taken is automatically recorded. The velocity at which the wave travels through the water is known from many previous measurements and so the distance travelled by the wave may be calculated. In practice, this is done electronically, and the instrument usually produces a visual trace of these variations in water depth.

The principle of seismic reflection surveying is precisely similar, except that the 'reflector' for which we are looking is not the junction between sea water and the loose sediment on the sea bed, but rather is the junction between two or more sets of beds. A reflection survey will normally be designed to record many different reflectors, usually major bedding planes lying at considerable depth. Therefore more penetration is required and so more powerful energy sources are needed than for the echo sounder. The use of a lower frequency source enhances the depth of penetration of the waves of energy.

Energy sources

At one time, explosives were virtually the only source of seismic energy, but there is now a considerable variety of energy sources available. Most of these are safer and cheaper and allow a more rapid 'rate of fire'. The main methods in use are summarised in Table 4.1.

TABLE 4.1 - ENERGY SOURCES IN SEISMIC REFLECTION SURVEY

ENERGY SOURCE	NATURE OF SOURCE	AREAS OF USE	APPLICATION
AIR GUN	Sudden release of highly compressed air	at sea	Structure of deep sedimentary layers for possible oil and gas traps
EXPLOSIVES	Explosion produced by specially developed geophysical explosives	land and sea	As above, but used less now, for safety reasons
'VIBROSEIS' (Trade mark of Continental Oil Company)	Continuous vibration produced by electro-mechanical means	on land	As above, used particularly in built-up areas to minimise disturbance.
SPARKER	Electric arc discharge, like a giant spark plug	at sea	Sediment on sea bed and at shallow depth below
BOOMER	Metal plate rapidly repulsed from electro-magnetic coil	at sea	As above

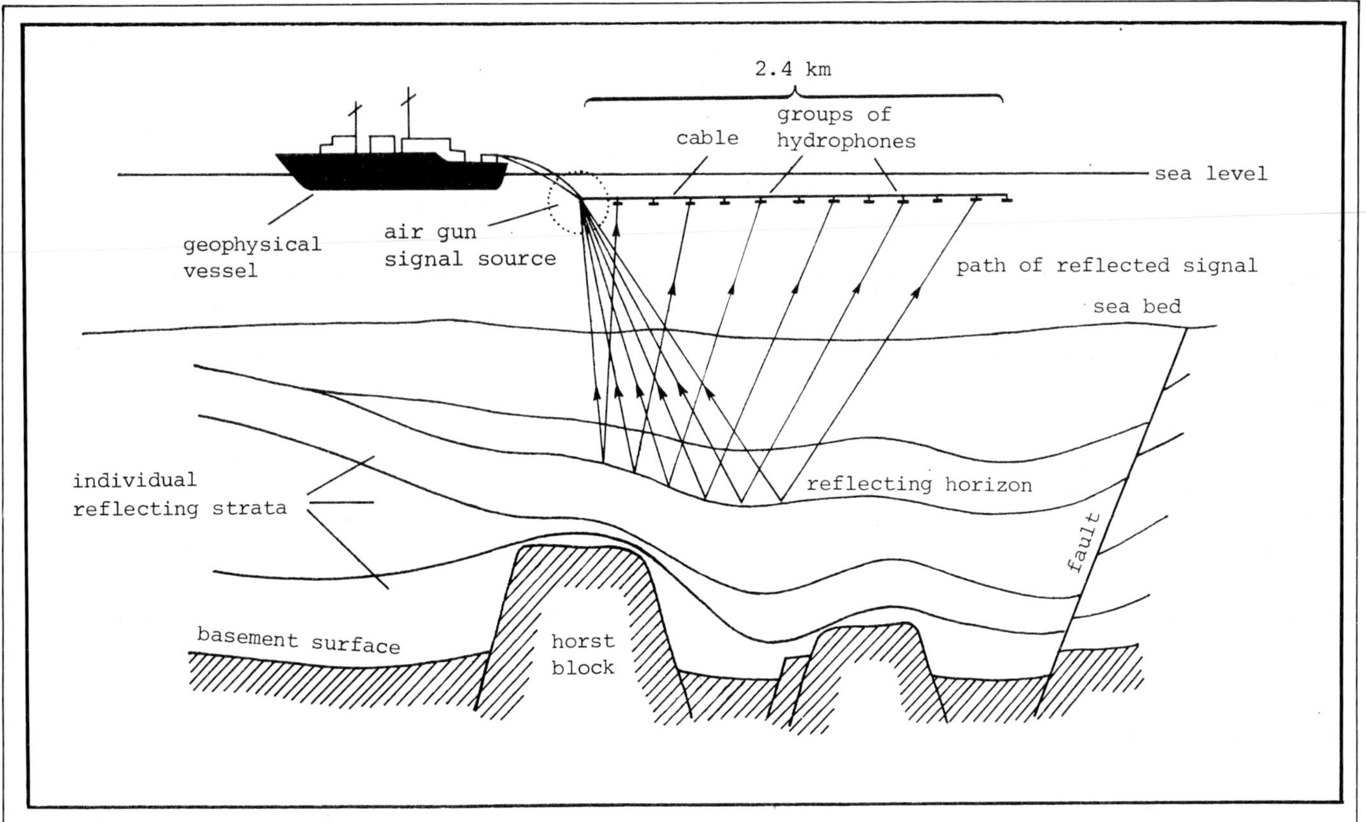

Fig. 4.1 Diagram of offshore seismic reflection method.

Detection equipment

Some of the energy which is reflected back to the surface is picked up by small detectors known as geophones (hydrophones at sea). These either employ a moving-coil or moving-magnet system or else a piezo-electric crystal. Before the advent of computers, 20 or so geophones would be embedded in the ground for each shot and the reflected 'arrivals' of energy picked off a paper trace 'by hand'. Most of the world's long-established oil-fields were first discovered by this means but since the early 1960s computers have been programed to facilitate the rapid analysis of data from many more geophones. For example, at sea a survey vessel tows both an air gun source and a long 'streamer' of hydrophones, inserted into a buoyant cable. The streamer may be as long as 2.4 km, containing 48 recording sections of cable, each one packed with a group of hydrophones connected in parallel (Fig. 4.1).

The vessel can steam continuously on a set course, 'firing' the air gun every few seconds and recording the arrivals of reflected energy on magnetic tape on board the ship itself. Many reflections are thus obtained from the same points on each bed beneath, and these are 'stacked' by the computer. This effectively reduces errors from false reflections, such as a wave which has bounced back and forth several times between the reflecting bed and the surface.

The final result is compiled by the computer and presented as a visual print-out.

North Sea oil and gas fields

The cover of this Unit shows a seismic reflection profile some 18 km long in the North Sea. Subsequent work proved the existence of a hydrocarbon field known as the Beryl Oil Field which is commercially viable. The seismic record clearly shows the anticlinal shape of the reflecting horizons, representing different strata below ground.

It should be noted that the visual record is presented as a sophisticated 'graph' of horizontal distance versus reflection time. It does not give the actual depth of each reflecting layer. This can only be done when we know the velocity at which the energy has travelled through each layer. In the first instance, these velocities are derived from advanced computer analysis of the data, or from specially placed extra shots. Subsequently, these may be confirmed by bore-hole logs, if the area is drilled.

Most of the natural gas and oil reservoirs of the North Sea have been located, since 1964, by the use of such methods. No geophysical techniques can actually prove the presence of gas or oil: only a borehole can do that. Even today, only one or two boreholes in ten actually find any oil or gas, but, nonetheless, the 'dry' holes make an invaluable contribution to the understanding of the structure and stratigraphy of a region. As an example, we can follow the fortunes of the holding company of the Beryl Field, Mobil North Sea Ltd.

An initial aeromagnetic survey was carried out over the North Sea, including Block 9/13, selected by the company. Fig. 4.2 shows the location of the block. The aeromagnetic map extract is printed as Fig. 3.6, with the limits of Block 9/13 outlined. There is a positive anomaly within the block, indicating that magnetic 'basement' rocks are nearer the surface here than in the immediate

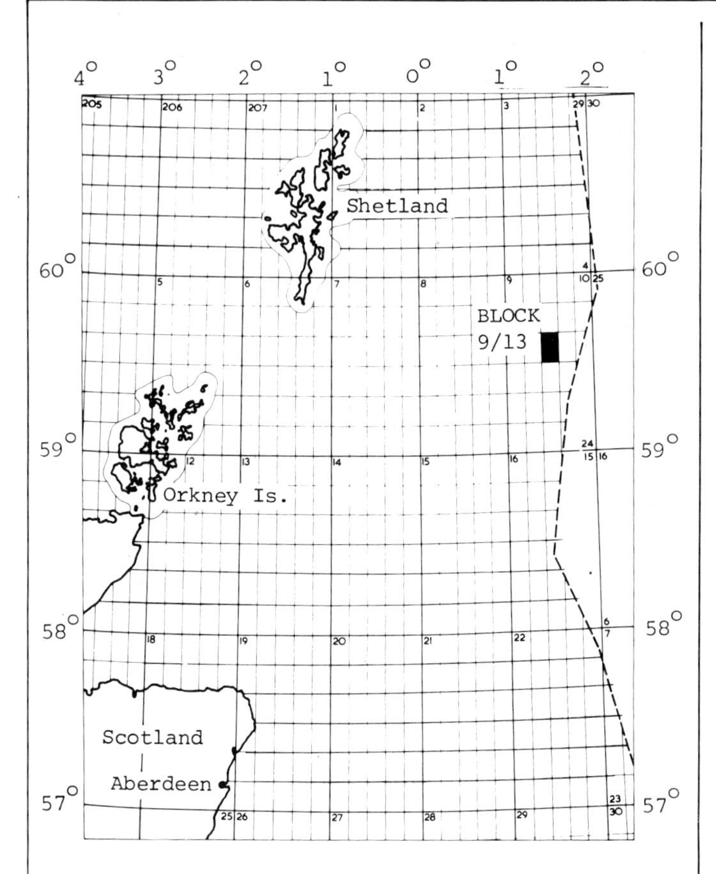

Fig. 4.2 Location of Block 9/13 in the North Sea. The dotted line is the limit of the British sector.

surroundings, possibly having been brought up by block faulting. Calculations of the approximate depth to the 'basement' may be made.

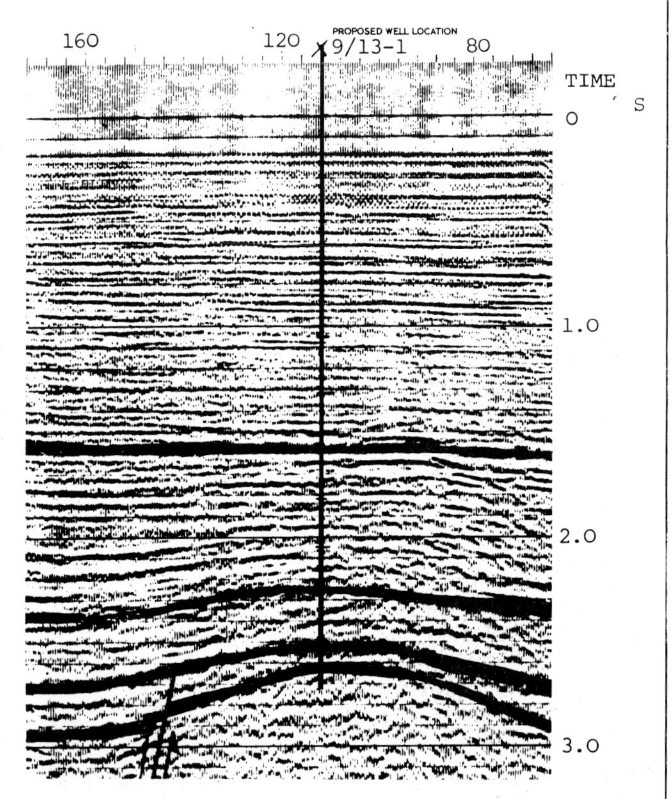

Fig. 4.3 Detailed seismic profile of the Beryl Prospect in Block 9/13. Solid lines have been drawn to emphasise major reflecting horizons and a fault. Numbers at the top are shot points. Times are for the double journey of the shock waves (outward and reflected).

The magnetic survey was followed by seismic reflection profiling, the results of which are depicted on the cover. Fig. 4.3 is a more detailed profile of the Beryl Field with possible geological boundaries sketched in and a fault inferred on the western side of the structure. This is as far as the geophysicist can go: the next step is for the geologist to provide the most likely geological interpretation regarding the ages and types of rock present. At first this was achieved by extension of the known geology of the adjacent land areas, but was rapidly improved once borehole programmes were begun in other parts of the North Sea. Figure 4.4 is a sketch of the expected geology of the Beryl structure, with the location of the first 'wild-cat' well.

PROPOSED WELL
LOCATION 9/13-1

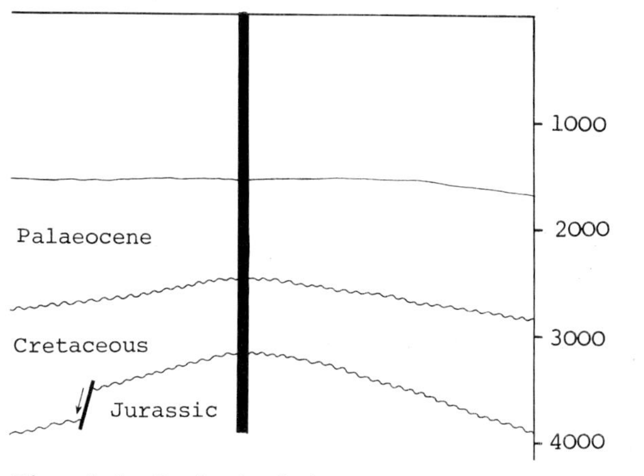

Fig. 4.4 Geological interpretation of the Beryl Prospect in Block 9/13. Length of section about 8 km. Intended total depth of exploration well 3810 m.

In this case, the painstaking work was rewarded. Almost 5 years after the project was planned the well struck oil at 3002 m, with more beneath. The well was stopped at 3148 m. Initial tests were favourable and a production platform was set up to exploit other parts of the field by further boreholes.

Even at this stage, some surprises may occur. Figure 4.5 shows the results of drilling a deflected well from a platform on the Dunlin Field, owned by the Esso Petroleum Company Ltd. The well logs show that the structure revealed by seismic work is not quite such a favourable reservoir for oil as had been thought.

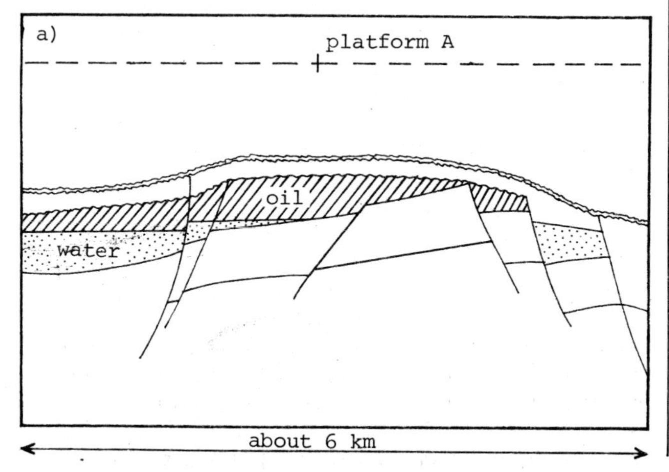

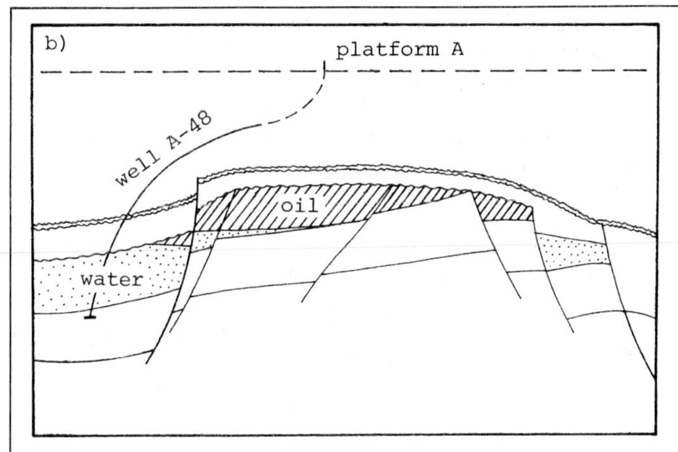

Fig. 4.5 The sectional diagrams show a) the way in which the oil and water deposits were assumed to be lying in the Dunlin reservoir and b) how the interpretation changed after well A-48 was drilled. This well revealed that the westernmost fault block was lower than had originally been thought.

Refraction techniques

At first sight, the principle of surveying by seismic refraction methods does not seem so obvious as the reflection technique, but in practice, refraction surveys are of great value. Shallow refraction work, using small explosive charges, a sledgehammer, or a 'thumper' (a heavy weight dropped from a lorry) is used in site surveys for foundation work in major civil engineering projects. Deeper crustal studies may also be made, and under favourable conditions, the entire thickness of the earth's crust may be determined by using charges of up to several hundred kg of explosives.

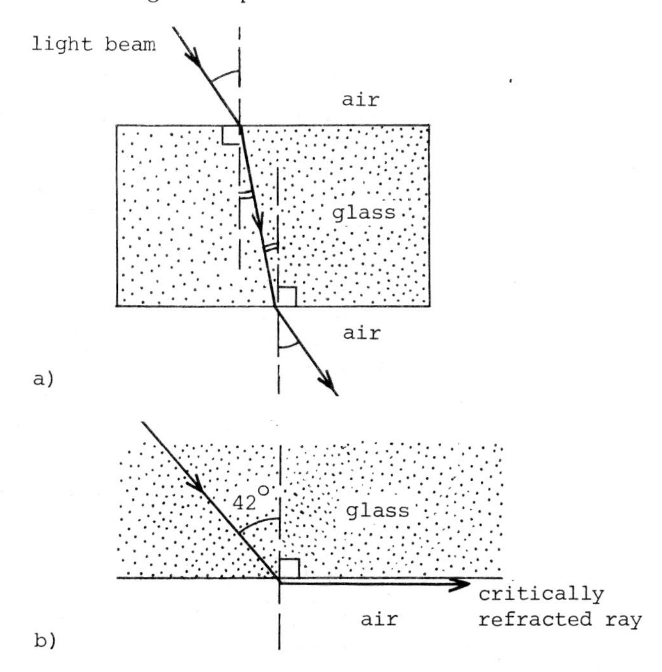

Fig. 4.6 Refraction of light through a glass block.

The principle of the refraction of waves of energy will be familiar to anyone who has experimented in school physics with a beam of light and a glass block. The light beam can be seen to bend as it enters and leaves the block (Fig. 4.6a). If the angle of incidence of the light beam is altered, a point is reached where it fails to emerge from the other side of the block; it has been refracted at the critical angle and then travels parallel to the side of the block. For a glass block in air, the critical angle is about 42°, depending on the type of glass (Fig. 4.6b).

A very similar thing happens in the earth when various layers occur, one on another. Different rock types transmit shock waves at different velocities, just as light travels more slowly in glass than it does in air. In nature, velocities at which shock waves are transmitted through rocks generally <u>increase</u> as the wave travels deeper. Where deeper strata are of lower 'seismic velocity' than those lying above, they are very difficult to detect.

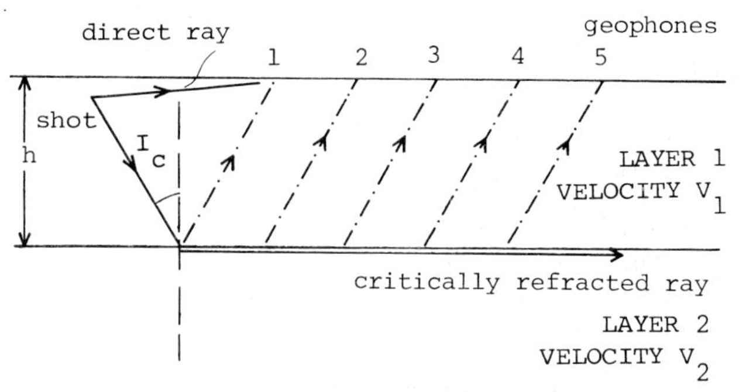

Fig. 4.7 Refraction of shock waves in the surface layers of the earth.

The basic principle is shown in Fig. 4.7. Waves from the explosion travel out in all directions, like the ripples on a pond when a stone is thrown in. The passage of the wave front in any direction is shown diagrammatically by an arrow. Some energy travels near to the ground surface (the direct wave), some is refracted into the next layer and the energy is eventually dissipated. One wave path, however, meets layer 2 at the critical angle (I_c) and is refracted along the top of layer 2, travelling at the new velocity V_2.

Energy is continually being returned into Layer 1 and this can be detected by the geophones. The time between the shot and the first arrival of the shock waves at the geophones may be measured very accurately (usually to better than a millisecond). At first, the direct wave will be the first to arrive, but as the geophones are spread further from the shot, the critically refracted wave will be received. Usually, V_2 is greater than V_1 and so eventually the refracted wave is the first to arrive, since it has covered most of its journey at the faster speed. This principle is continued with arrivals from deeper layers. Arrival times are plotted against the distance between the shot and the geophones (Fig. 4.8).

Such a graph is known as a 'travel-time plot' and the velocity of seismic wave transmission in each layer is easily obtained from the graph. In the form in which it is plotted in Fig. 4.8, V_1 is obtained from the reciprocal of the gradient of the line XY and V_2 similarly from the line YZ. Try measuring V_1 and V_2 for yourself.

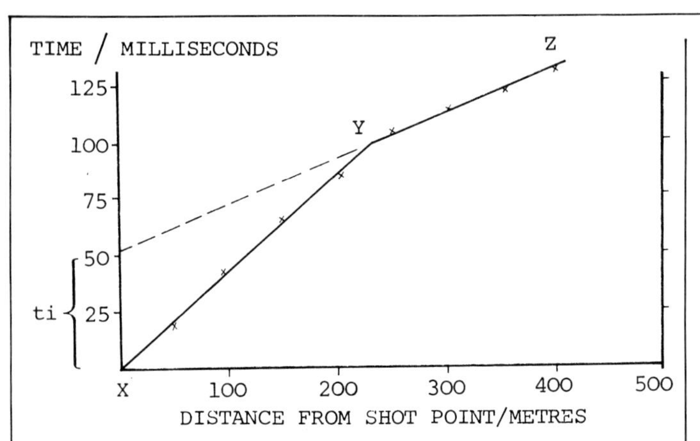

Fig. 4.8 'Travel-time' plot for a two-layer seismic refraction profile.

'Outcrop shooting' on rocks of all types has enabled us to determine characteristic (but not unique) values for seismic wave velocities. Some typical figures are:

Dry sand/gravel	500-1000 m s^{-1}
Shale	1000-3000 "
Sandstone	2000-4300 "
Rock Salt	4700-5700 "
Slate	4700-5300 "
Basic igneous rocks	up to 7700 "

Which of these rocks types could be present for the survey shown in Fig. 4.8?

In addition to the velocities in the layers, it is also possible to calculate the layer thicknesses. The appropriate equation for Layer 1 (derived ultimately from Snell's Laws of Refraction) is:

$$\text{Thickness } h = \frac{ti}{2} \frac{V_2 V_1}{\sqrt{V_2^2 - V_1^2}}$$

where ti is the 'intercept time' in seconds (i.e. 0.052 seconds in Fig. 4.8)

Calculate the thickness of Layer 1. You cannot of course deal with Layer 2 until the travel-time curve indicates another change in velocity as the next layer down is reached.

Two examples of the application of the seismic refraction method illustrate the versatility and some of the uncertainties of the technique.

Hammer seismic profile

A seismic line was 'shot' across the inside of a meander bend by a party of sixth formers, the equipment used being a simple seismic set with one geophone. The shock-wave was produced by the most muscular student striking an iron plate with a sledge hammer. The moment of impact was recorded from an impact-switch attached to the hammer, thus enabling travel times of the wave to the geophone to be electronically recorded. A sketch map of the locality and the travel-time plot is shown in Fig. 4.9.

Compute V_1 and V_2 and the thickness of Layer 1. Then attempt your own brief geological interpretation, using, in addition, the data shown on the sketch map. Assume the refractor is horizontal. A suggested answer is given at the end of the Unit.

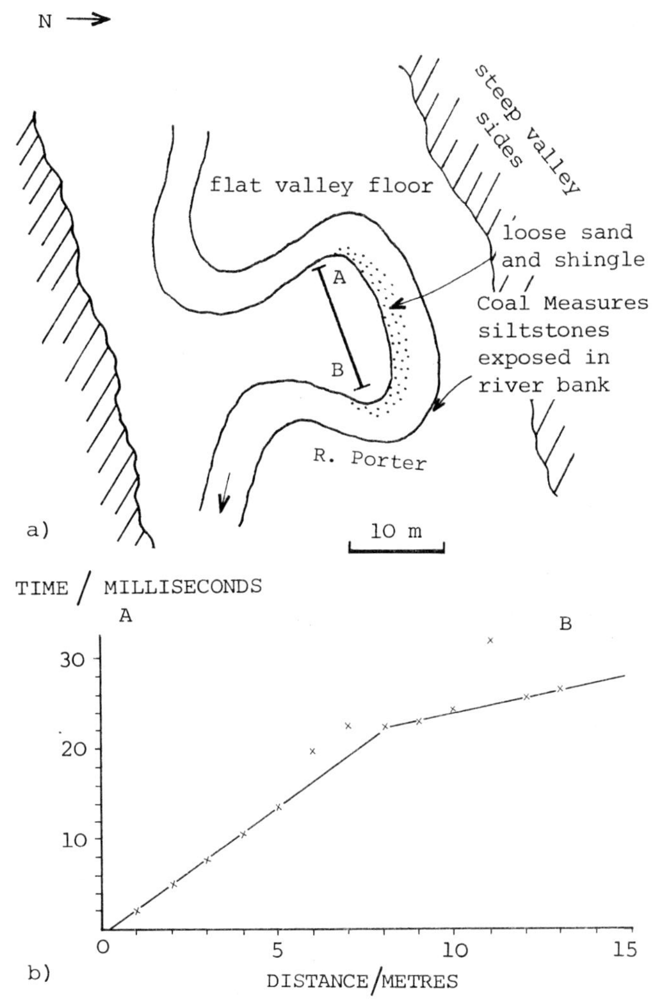

Fig. 4.9 a) Sketch map of a section of the River Porter, Sheffield (SK 312852) showing location of seismic refraction profile AB. b) Travel-time plot for the line AB.

Seismic refraction survey of Cardigan Bay, Wales

During the 1960s, the University of Birmingham carried out a research project involving seismic refraction and 'sparker' surveys from ships in Cardigan Bay. The survey was initiated to try to solve an interesting problem.

Earlier gravity surveys of West Wales had shown a rapidly decreasing Bouguer anomaly towards the coast (see Fig. 4.10). This was assumed to be due to low density rocks lying along the coast and beneath the Bay and the gravity gradient was so steep as to suggest a faulted boundary parallel to the coast. What could be the cause of the gravity 'low'? Reference to a copy of the 10 mile map of the country will show that much of the geology of West Wales comprises Lower Palaeozoic rocks, mostly greywackes, shales and slates. Large granitic bodies are absent: so too are major deposits of Mesozoic or younger rocks, either of which could reasonably be expected to provide the necessary density contrast. Most of the palaeogeographic maps available

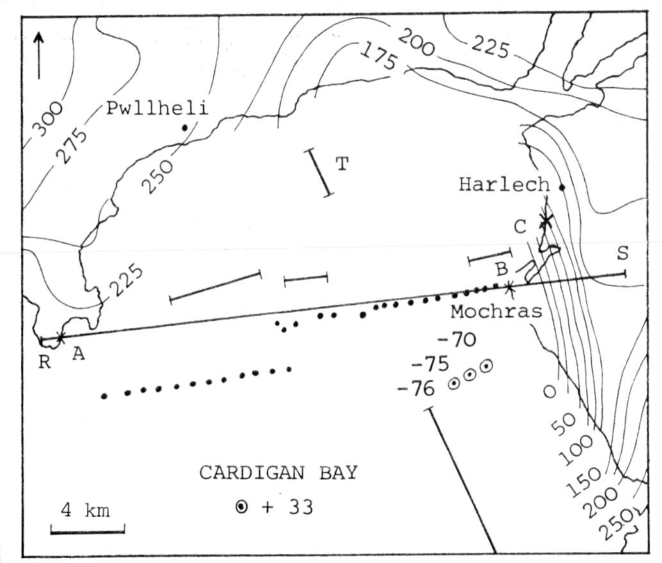

KEY — 10 Bouguer anomaly in g.u.
⊙ gravity station on reef Bouguer anomaly in g.u.
├──┤ sea seismic line
• shots to land seismic stations A, B, C.

Fig. 4.10 Gravity anomalies and seismic stations in part of Cardigan Bay.

in the 1960s interpreted West Wales and Cardigan Bay as having been land areas from the end of the Silurian to the Cretaceous (e.g. Fig. 4.14a).

On the basis of this brief survey and the 10 mile geological map, can you suggest possible causes for the trend of the gravity anomalies, before you read further?

The seismic refraction survey was carried out in the northern parts of the Bay and was later followed by sparker surveys to delimit the near-surface sediment. Where possible, each refraction line was 'reversed' to eliminate errors due to the dip of the strata. A typical travel-time graph is shown in Fig. 4.11.

The results of several seasons' work are summarised in the interpretation in Fig. 4.12. This is as far as direct interpretation

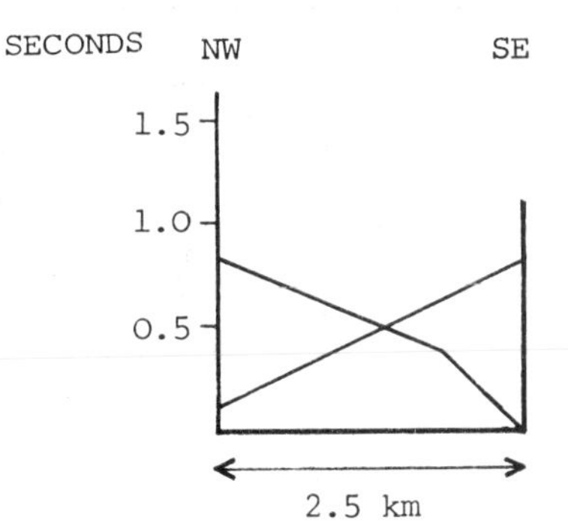

Fig. 4.11 Travel-time graph for line T.

of the seismic results will allow; the geological meaning of the profiles is open to discussion and, as is usual with such work, the research workers presented their findings to a learned society in two papers.

Table 4.2 summarises the ideas put forward by the authors and some of the suggestions made by other speakers in the discussion which followed the lectures.

Clearly, there was considerable diversity of opinion, much of it coming from closely reasoned geological arguments from geologists well versed in the rocks of Wales. The Birmingham group's second paper concluded with the comment that a borehole at Mochras, on the westernmost part of the coast would clinch the argument.

A few years later, just such a borehole was undertaken. The results (Fig. 4.13) were quite a surprise. It now seems that much of 'Layer 1' is of middle Tertiary age, a time which is sparsely represented in the rocks of mainland Britain! Furthermore, Layer 2 corresponds to the most complete succession of Liassic (Lower Jurassic) rocks anywhere in the country!

The implications of the combined geophysical and borehole studies are far reaching. First, sedimentary basins of Mesozoic and Tertiary age are usually the most likely reservoir rocks for oil and gas. Commercial exploration for oil and gas is now being carried out in the seas west of Britain.

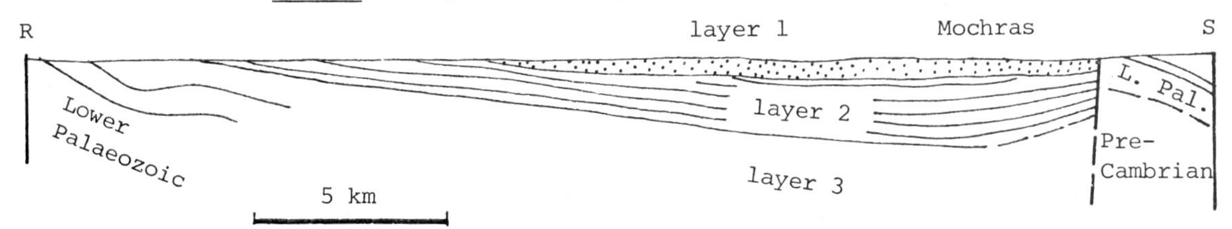

Fig. 4.12 Interpretation of the structure of part of Cardigan Bay from geophysical data.

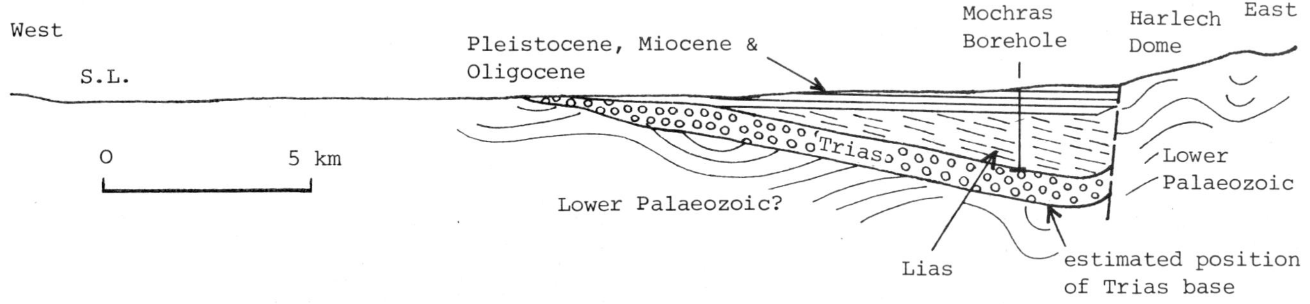

Fig. 4.13 Section through Tremadoc Basin and Mochras boring.

TABLE 4.2 - GEOLOGICAL INTERPRETATIONS OF GEOPHYSICAL WORK IN CARDIGAN BAY

PROBLEM	SOLUTIONS SUGGESTED BY BIRMINGHAM UNIVERSITY GROUP	DISCUSSION AT TWO MEETINGS OF THE GEOLOGICAL SOCIETY OF LONDON
NATURE AND AGE OF LAYER 1	Post-glacial material overlying Tertiary and/or Mesozoic clays. (Jurassic fossils occur in beach pebbles. Tertiary clays crop out in Northern Ireland)	a) Whole of Layer 1 could be glacial material, by analogy with Scottish example (K Dyer) b) Could be Trias, similar to Midlands (T N George). c) New aeromagnetic survey shows few anomalies in Cardigan Bay. Could be Trias or Tertiary (W Bullerwell).
NATURE AND AGE OF LAYER 2	a) Ordovician sediments. (Known on mainland. Seismic velocities consistent with outcrop shooting on Ordovician) b) Could be Permo-Trias or Carboniferous, but unlikely.	a) Carboniferous, on palaeogeographic grounds. (T N George) b) Trias - similar to other Triassic basins in Cheshire (O T Jones). c) If junction between Layers 2 and 3 is really the base of the Ordovician, why is it not folded, as on the mainland (W F Whittard)?
NATURE AND AGE OF LAYER 3	Cambrian (known on mainland).	No comment.
NATURE OF EASTERN BOUNDARY OF CARDIGAN BAY	Major fault (from steep gravity gradient and seismic work on land).	Not faulted. Prefer an overlap junction (T N George).
GENERAL STRUCTURAL SETTING OF CARDIGAN BAY	a) Structural basins preferred. b) Buried granite beneath surface sediment. Top of granite would be at 7 km. depth and base at 26 km. Possible from gravity anomaly but unlikely.	a) New sea-borne gravity profile shows 60 g.u. negative anomaly. Consistent with deep sedimentary basin <u>or</u> buried granite (M H P Bott). b) Could be late Mesozoic/Tertiary downwarp as known in western English Channel (S E Hollingworth).

Secondly, our ideas of the palaeogeography of the country at various times in the past must be completely redrawn. Figure 4.14a shows the previously proposed palaeogeography for the Lower Jurassic and Fig. 4.14b a more recently drawn one in the light of offshore work such as that described above. Quite a lot of land seems to have sunk since 1951!

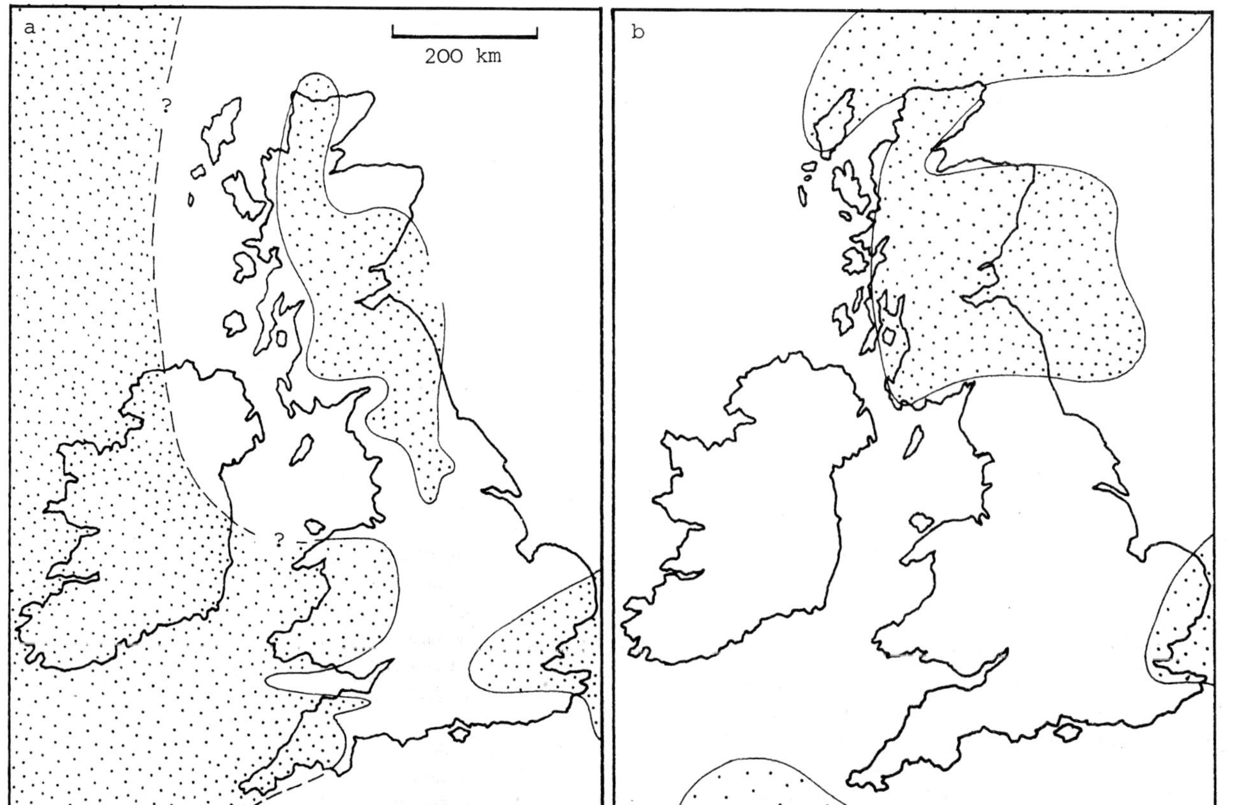

Fig. 4.14 Two interpretations of the palaeogeography of the British Isles during Lower Jurassic times a) after L J Wills 1951 b) after A Hallam and B W Sellwood 1976. Land areas stippled.

Earthquake Seismology

General principles and wave types

Broadly speaking, the principles involved in employing earthquake energy as a seismic source are similar to those of explosion seismology. Shock waves are sent out in all directions following an earthquake and they are refracted as they pass through layers with different seismic velocity characteristics. Because the natural energy produced is so much greater than that from man-made sources, waves from a strong earthquake actually travel right through the earth and can therefore provide information about the deep interior. The time of arrival of the shock wave as it emerges after its journey is recorded on a seismograph, which acts in the same manner as an elaborate geophone. The seismograph converts the shock wave arrivals to electrical impulses which, in turn, are recorded as 'wiggly line' traces on a moving-drum recorder. Accurate time marks are also made on the recorder, so that the precise time of arrival of each seismic 'event' can be read later.

To make full use of earthquake seismology, we need to study more closely the nature of the shock waves. For most straightforward purposes in seismic surveying, using man-made sources, we can regard the waves as being of one type, but this is not true of earthquake seismology. Earthquakes produce several different types of wave, which travel in different ways and with different velocities.

Body waves (P and S waves)

Those which travel through the earth are known as 'body-waves' and comprise two types. The faster-travelling ones are known as P-waves (or longitudinal waves) and they are propagated by a series of compressions and rarefactions of the particles in the earth. In a way, they may be likened to a string of railway trucks being shunted. As the engine hits the first truck, the movement is passed on to the second truck, but the first one rebounds. This continues down the line until the furthest truck moves. After the train gets under way the analogy is no longer valid, since all the wagons are moving together and not just oscillating backwards and forwards over one spot.

The slower body waves are known as S waves (or transverse waves). These are transmitted through a medium by the motion of particles at right angles to the direction in which the wave is travelling.

Surface waves (L Waves)

Two other types of wave are also produced. These do not travel deep in the earth, but are restricted to the surface layers. Such surface waves are known collectively as L waves and comprise Love waves which are transverse waves and Rayleigh waves, which travel with an elliptical particle motion. L waves are capable of being transmitted right round the globe and it is these waves which are responsible for the ground-roll which produces such devastation during an earthquake. The P, S and L waves appear on the seismograph record as shown in Fig. 4.15.

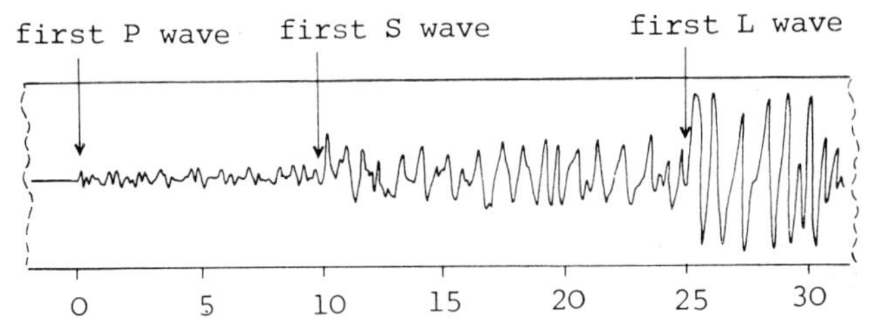

Fig. 4.15 Part of seismogram for an earthquake some 8500 km from the recording station.

The knowledge of wave types is of immense practical interest for several reasons. First, they can be used to enable the distance between the earthquake and the observatory to be calculated. There is fortunately an empirical relationship between the time of arrival of the P, S and L waves and the distance from the source of energy. This has been plotted from 'known' earthquakes and is shown in Fig. 4.16.

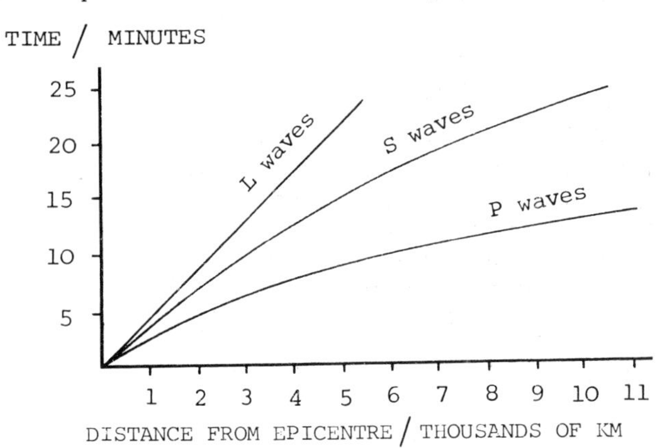

Fig. 4.16 Time-distance graph for P, S and L waves.

Calculations from the records of the same earthquake made at three or more seismograph stations enable the position of it to be pinpointed. The position may be plotted on a globe as the epicentre (that is, the point on the earth's surface directly above the actual centre or focus of the event itself).

Secondly, passage of the P and S waves through the earth depends upon the elastic properties of the medium through which the waves are passing. The relationship may be measured in the laboratory under controlled conditions and then applied to the earth's interior. Thus, P wave transmission is related to the properties of the medium as follows:

$$\text{P wave velocity} = \sqrt{\frac{K + \frac{4}{3}\mu}{\rho}}$$

Where K is the bulk modulus, or incompressibility of the medium; μ is its rigidity modulus (related to its shear strength) and

ρ is its density. The corresponding equation for S waves is:

$$\text{S wave velocity} = \sqrt{\frac{\mu}{\rho}}$$

Given that K is always positive, and that μ is zero for a liquid, try to answer the following:

1. *Which is always greater, P wave or S wave velocity?*
2. *Which type of wave will not be transmitted by a liquid?*
3. *How does elastic wave velocity depend on density?*

Your answer to 3 may have surprised you, since you probably expected a directly proportional relationship, not an inverse one. In fact, densities do increase with depth and so do wave velocities, but this is because the elastic moduli increase faster and more than counteract the rise in density.

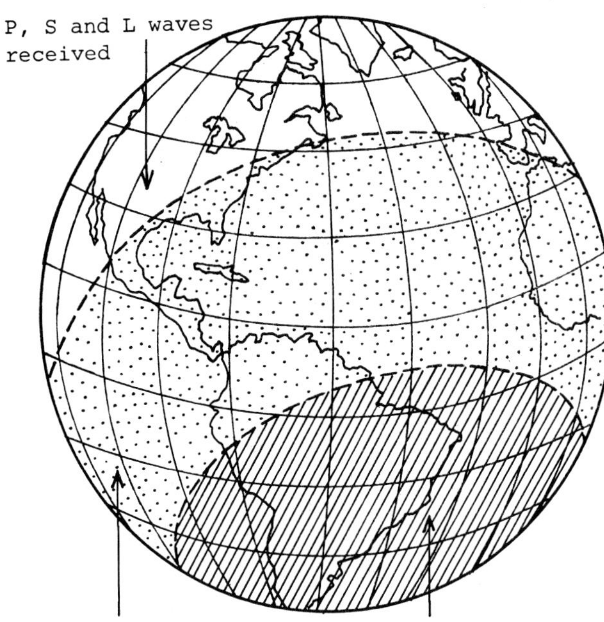

Fig. 4.17a The 'shadow zone' from an earthquake in Japan.

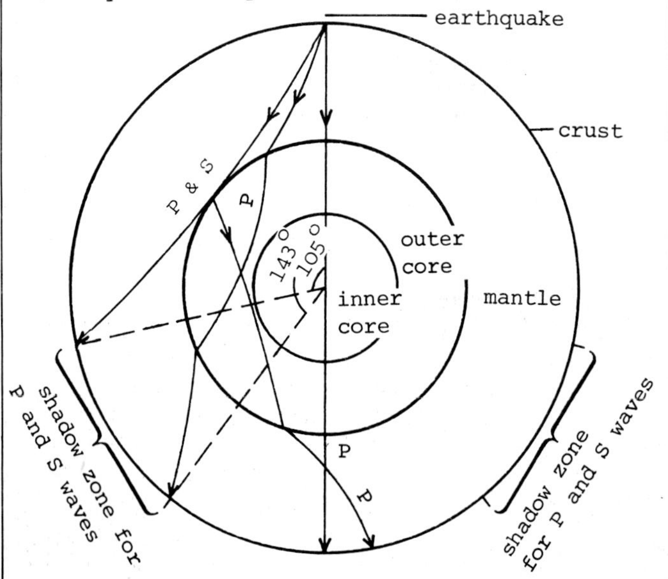

Fig. 4.17b The paths of selected P and S waves through the earth.

The deep structure of the earth

When the arrivals from many different earthquakes are plotted, some rather surprising results emerge. From any one earthquake, stations in some parts of the world record complete sets of P, S and L waves; others record P and L waves only, whilst others still receive only L waves. The distribution is not haphazard, but follows the pattern shown in Fig. 4.17a. The area where only L waves are recorded is known as 'shadow zone' and spans the earth, at an angular distance from the earthquake of between 105° and 143°. Beyond this zone, P waves are received again, but no S waves.

Our knowledge of the elastic moduli leads us to infer that at least part of the earth's interior must be liquid and that the S waves have been absorbed. The P waves have emerged after being progressively refracted on their passage through the earth. Figure 4.17b shows an interpretation of the wave paths through the earth and Fig. 4.18 is a plot of depth from the surface against wave velocity. There is a marked drop in the velocities at 2900 km depth and this is taken to mark the boundary between the mantle and the liquid core. The boundary is known as the Weichert-Gutenberg Discontinuity, after its discoverers.

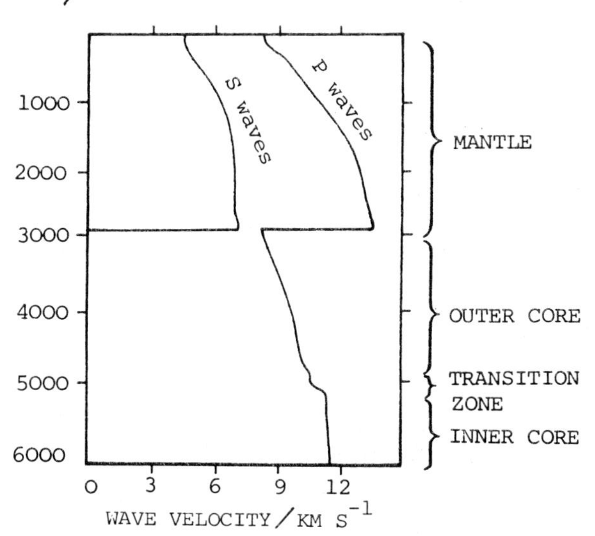

Fig. 4.18 Velocity/depth curves from the earth's surface to its centre (details of the crust omitted).

Deeper within the core, the P wave velocities begin to rise again. Theoretical considerations of the likely properties of materials at such great depths lead us to believe that this inner core is solid.

The nature of the crust from earthquake and explosion seismology

The thickness of the crust

Evidence for the nature of the uppermost layers of the earth mostly comes from near-earthquakes, which can be regarded in a similar way to man-made explosions.

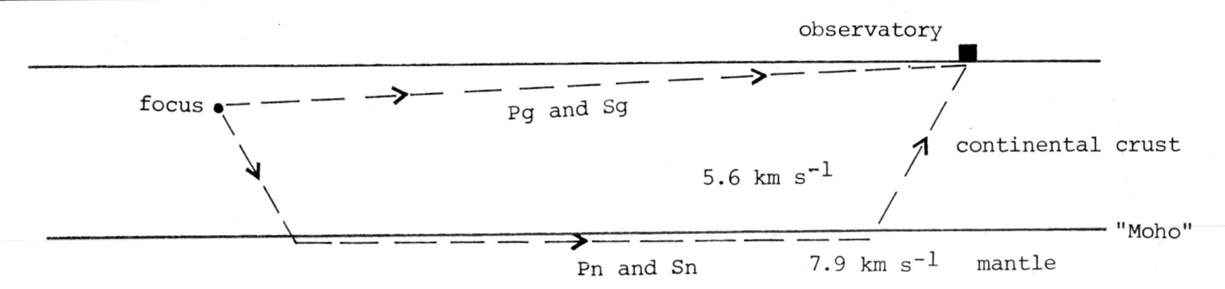

Fig. 4.19 Crustal structure determined from near earthquake.

A Yugoslavian named Mohorovičić was the first to notice two sets of seismic waves arriving at the seismographs from a single nearby earthquake shock which took place in 1909. He realised that the first waves to arrive represented energy which had been refracted at a depth of over 50 km below the surface and had travelled most of the journey in a higher velocity medium (P_n and S_n in Fig. 4.19). The second arrivals were produced by slower ground waves which had travelled direct to the observatories through the crustal rocks (P_g and S_g in Fig. 4.19). The refractor responsible seemed to represent a sharp junction separating the rocks of the crust from the underlying mantle and it soon became known as the 'Moho', or 'M-Discontinuity' after its discoverer. He calculated an average P wave velocity for the continental crust of Yugoslavia of 5.6 km s^{-1} and for the upper mantle beneath of 7.9 km s^{-1}.

Subsequent work in many parts of the world has proved the widespread occurrence of the M-Discontinuity, although it seems that in some areas it represents a gradual transition zone rather than an abrupt break. The seismic velocity of the uppermost mantle also varies, but, apart from some anomalous areas, its minimum value is 7.7 km s^{-1} for P waves, which is appreciably higher than that of most crustal rocks.

Near-earthquake and large-scale explosion seismology have also been applied to test the hypotheses of Airy and Pratt regarding the variations of thickness and density of the crust (see p. 8). It soon became clear that there are major contrasts between the oceanic crust and that of the continents. Such seismic work has shown that, as Airy predicted, the oceanic crust is both thinner and denser than the continental crust, the average thicknesses being 11 to 12 km (inclusive of sea water) and 35 km respectively.

Beneath many mountainous areas, the crust is usually much thicker than average. For example, for the Rocky Mountains it is over 60 km thick. This also seems to vindicate Airy's hypothesis, but more recent seismic work has shown that there are some regions, such as the Basin and Range Province of the western USA, where the crust is continental in type and yet only 25-30 km thick. Here, the observed gravity anomalies are more readily explained by lateral changes in rock density, as invoked by Pratt.

Structure of the oceanic crust

Not only the thickness, but also much of the detailed structure of the crust can be determined by seismic means.

The oceanic crust is rather less variable than that of the continents. It is also less troublesome to detonate large explosions at sea rather than on land. (Even so, 150 kg of TNT may be required to produce measurable refractions from the M-Discontinuity at sea.) Fig. 4.20 gives a typical result from refrac-

TABLE 4.3 - THE OCEANIC CRUSTAL LAYERS

LAYER	P WAVE VELOCITY (KM S^{-1})	AVERAGE THICKNESS (KM)
Sea Water	1.5	4.5
Layer 1	1.6 - 2.5	0.4
Layer 2	4.0 - 6.0	1.5
Layer 3	6.4 - 7.0	5.0
Upper Mantle	7.4 - 8.6	

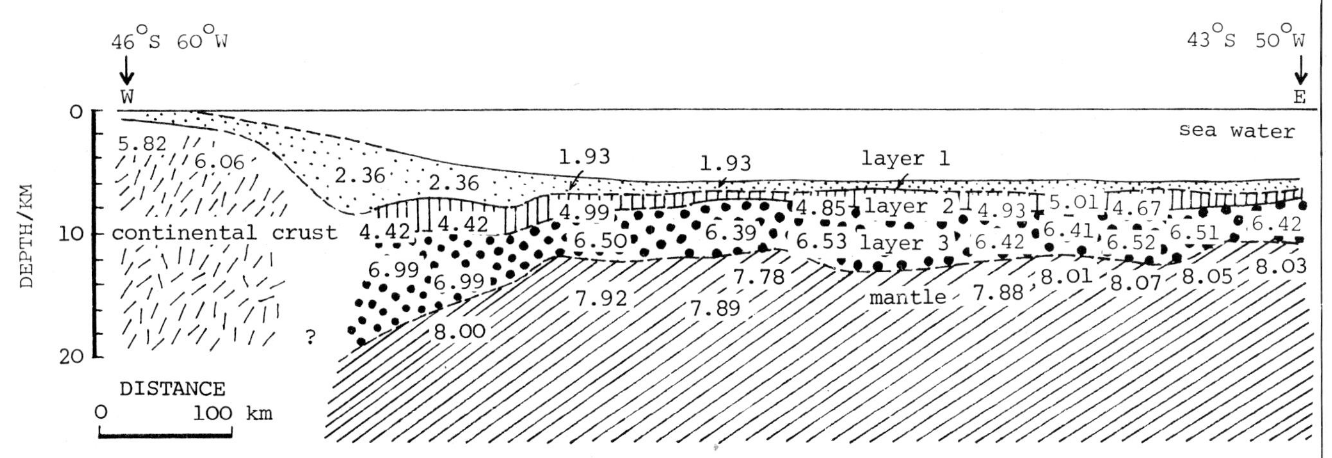

Fig. 4.20 Oceanic crustal structure determined by seismic refraction in the Atlantic ocean east of Argentina. Figures refer to seismic velocities in km s-1.

tion surveys in the South Atlantic Ocean. The compilation in Table 4.3 is a section through an 'average' oceanic crust, derived from many such surveys in all the major oceans. Seismic reflection work is also used to provide additional detail of the nature of Layer 1.

Geologically, the following interpretation seems most likely:

Layer 1: unconsolidated sediments passing downwards into consolidated sediments.

Layer 2: basalts, or consolidated sediments, probably the former.

Layer 3: partly metamorphosed basic igneous rocks.

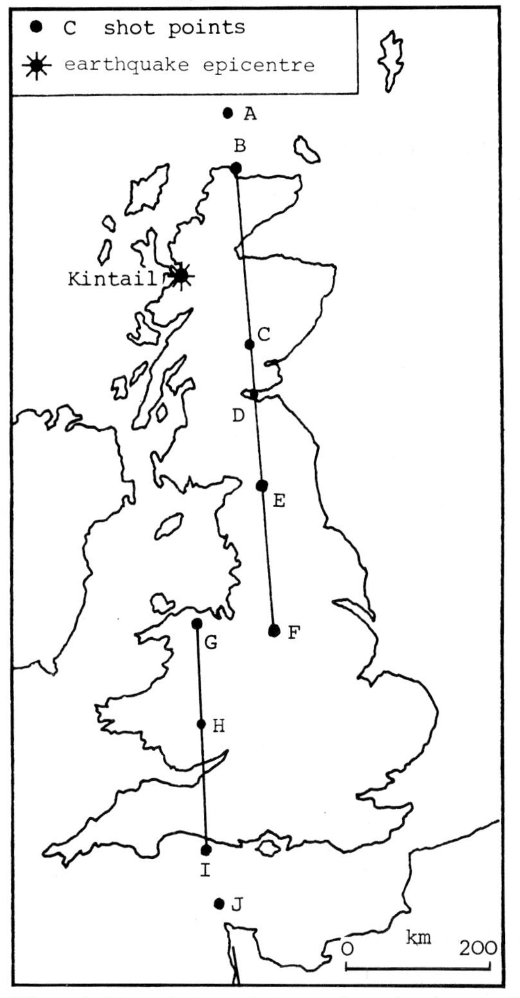

Fig. 4.21 Shot points and seismic profiles used during the LISB experiments of 1974.

Structure of the continental crust

Although it is more difficult to conduct deep seismic experiments on the continental crust, many have been undertaken in recent years. An example is the Lithospheric Seismic Profile in Britain (LISB) of 1974, where refraction lines were shot from the North of Scotland to the English Channel and across into Northern France (Fig. 4.21). A typical velocity-depth interpretation of the crust below the Scottish Highlands is shown in Fig. 4.22. In this case, the survey team were lucky enough to record a minor earthquake which occurred below the coast of West Scotland whilst the equipment was running.

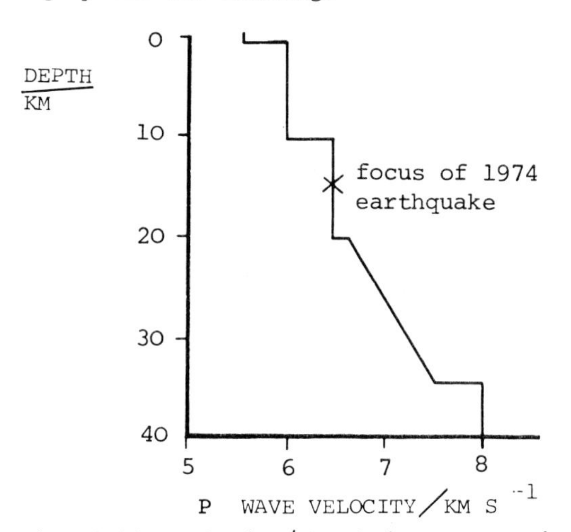

Fig. 4.22 Velocity/depth interpretation of data from the Kintail earthquake 1974.

A simplified version of the LISB team's interpretation of the depth of the M-Discontinuity beneath Britain is shown in Fig. 4.23. Much of it is close to the average depth beneath the continents of 35 km but there are significant variations. Compare this thickness with the average 11 to 12 km for the depth of the M-Discontinuity beneath the oceanic crust (Table 4.3).

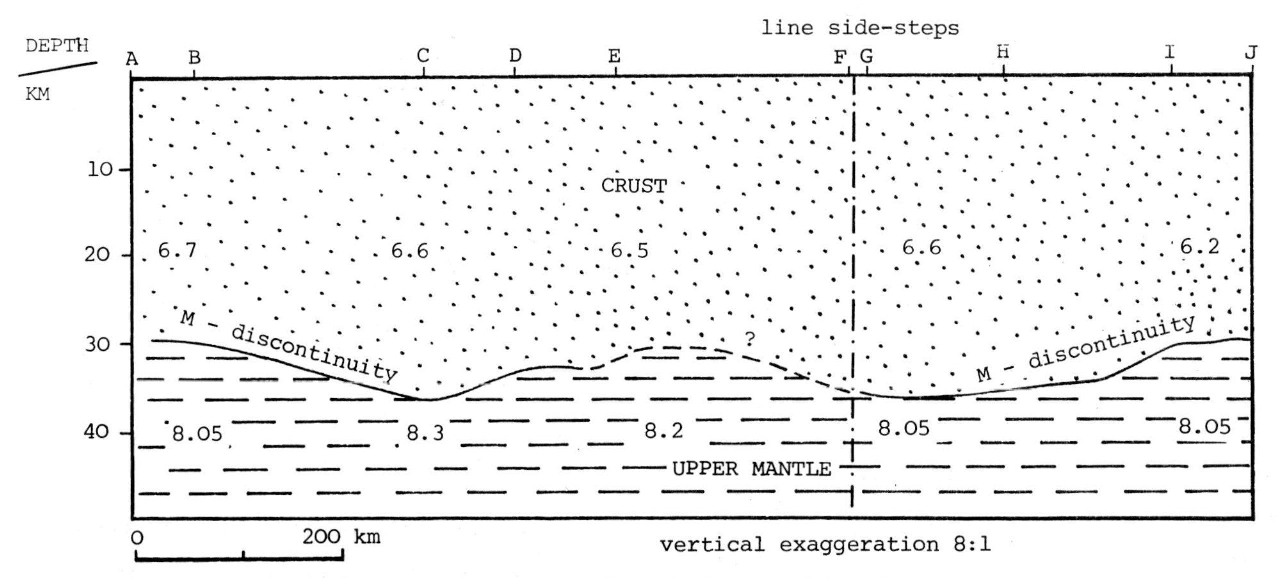

Fig. 4.23 Generalised crust-upper mantle section of the British Isles along the line A-J. P wave velocities are average figures for the whole crust and for the upper mantle and are in Km s-1.

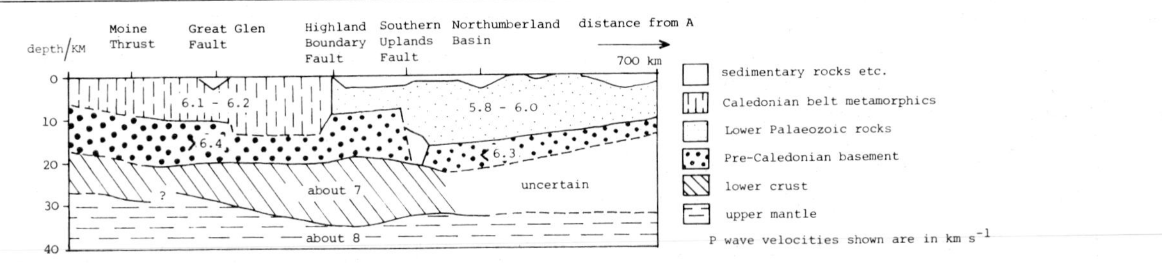

Fig. 4.24 Diagrammatic cross-section of the crust and upper mantle of northern Britain.

The LISB experiment has also resulted in a tentative 'model' for structures within the crust of Northern Britain, a simplified version of which is shown in Fig. 4.24. Several layers may be equated with the known sedimentary, igneous and metamorphic geology represented at outcrop. Beneath them lies the 'lower crust' which is of comparable velocity to Layer 3 beneath the oceans. Many workers believe that the geology of this 'lower crust' is broadly similar to that of Layer 3.

At one time, it was thought that a marked break in seismic velocities between an upper and a lower continental crust was always present. Indeed, in places where it could be measured the junction was referred to as the 'Conrad Discontinuity'. It is now clear, however, that there are very many regions where there is no sharp break within the continental crust and so the term is of doubtful validity.

The low velocity layer

Recently, careful examination of seismograph records has revealed the existence of a layer, or zone, within the upper mantle, where seismic velocities are actually lower than those of the overlying layers (Fig. 4.25). This is usually taken to mean that the layer must be physically weak or unusually plastic, and it is sometimes given the Greek name 'asthenosphere' to express this fact.

By contrast, the parts of the mantle lying above the asthenosphere together with the overlying crust, form a relatively brittle shell, known as the 'lithosphere', or 'rocky shell'. On average, the lithosphere is between about 100 km and 200 km thick. For many purposes, we can regard the crust and upper mantle which comprise the lithosphere as acting in unison, able to 'ride' as it were, over the weak asthenosphere.

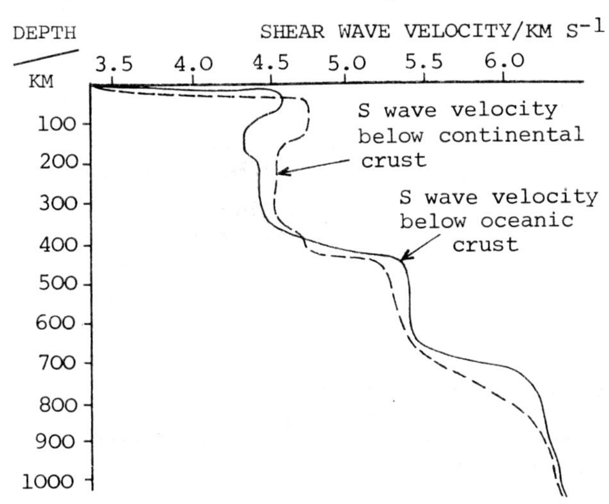

Fig. 4.25 The low velocity layer.

The surface distribution of earthquakes

So far we have only considered the information yielded by earthquake waves travelling through the earth, but there is also much to be learned from a study of the surface distribution of earthquakes. It is common knowledge that some parts of the world are more subject to earthquakes than others, although nowhere can be considered to be totally immune. The map (Fig. 4.26) shows the distribution of the main earthquakes and it is immediately clear just how restricted in occurrence they are.

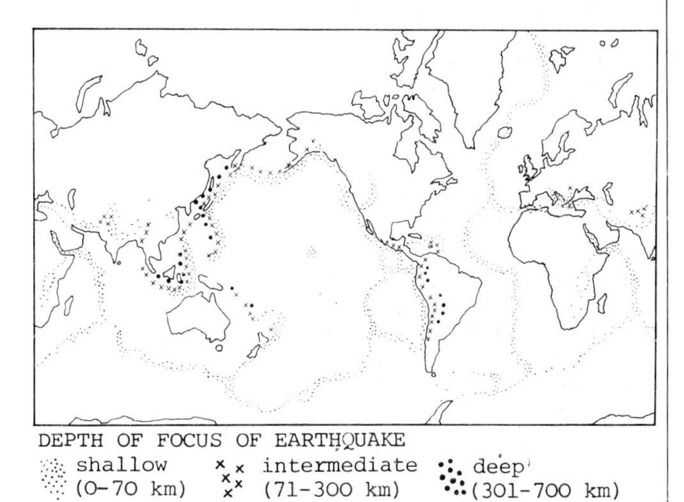

Fig. 4.26 The distribution of earthquakes.

The most marked seismic belt runs three-quarters of the way around the margins of the Pacific Ocean. A similar line bisects the Mediterranean area and continues on through southern Asia and Indonesia to meet the circum-Pacific belt. Smaller belts occur in the Caribbean and Scotia Seas. Narrow belts of earthquakes run through the major oceans of the world, whilst the East African Rift Valley is marked by a further line of seismic activity. Other, minor earthquakes occur in association with isolated volcanoes such as the Hawaiian islands.

Noticeable though all these earthquake epicentres are, it must not be overlooked that by far the greatest part of the earth's surface is largely aseismic, that is, inactive. Why then are the earthquake zones so special? When we examine them more closely, we find that each major type of seismically active belt corresponds with a surprising number of other geological features. In list form these are as follows:

Circum-Pacific belt, Caribbean and Scotia Sea

1. Earthquake foci lie along a plane often inclined at about 45° to the horizontal.

In regions such as the Pacific coast of South America, shallow foci (0 to 30 km in depth) are on the seaward side. Intermediate ones (71 to 300 km in depth) lie roughly beneath the coastal areas, whilst deep focus events (301 to 700 km in depth) lie beneath more inland areas (Fig. 4.26).

2. Where the earthquake belt lies near to the junction between a continent and an ocean, the continental margin usually contains a young fold mountain range, frequently with active volcanoes.
3. Other parts of the belt, in the open ocean, are marked by island arcs, that is, chains of mostly volcanic islands arranged in a gentle curve, whose convex side is towards the middle of the ocean.
4. The dominant lava type in the island arcs and fold mountain ranges is andesite, a lava of intermediate composition.
5. A deep trench in the ocean floor usually lies off the convex side of the island arcs or off the coast of the fold mountain ranges.
6. Markedly negative gravity anomalies occur in association with these regions.
7. Heat flow outwards from the earth in such areas is abnormally high, i.e. there is a heat surplus being generated by some process deep in the earthquake zone.

Some of these observations also apply to the Mediterranean/Himalayan earthquake belt.

Mid-oceanic earthquake belts

1. Earthquakes are of shallow depth only (0 to 70 km depth).
2. The lines of earthquakes lie beneath major ocean ridges, which rise like underwater mountain chains some 2000 to 3000 metres above the general level of the sea floor.
3. The ridges are associated with vulcanicity of basaltic lava type.
4. Heat flow over the oceanic ridges is abnormally high.
5. The rocks of the sea bed exhibit magnetic 'sea floor stripes' as described in an earlier section.

Plate tectonics – a brief note.

We have already attempted to explain some of the features of the ocean ridges. Presumably the shallow earthquakes with which they are associated result from the generation of magma in the upper mantle and the 'pulling apart' of the lithosphere as the magma is injected or extruded. If this process alone were acting throughout the world, it would be necessary for the earth to be continually expanding to accommodate the new lithospheric material, but few geologists believe this to be the case. Instead, it now seems that we may look at the other major earthquake zones, such as the circum-Pacific belt, as the site of a kind of 'return mechanism' whereby at least some of the lithospheric rocks are returned to the mantle. The angle from the horizontal at which the earthquake foci are inclined, and the preponderance of features of compressional origin, suggest that one portion of lithosphere is being pushed down against another, e.g. where a continent meets an ocean, it is the material on the oceanic side which is pushed down (Fig. 4.27).

Whilst we have concentrated on all the exciting activity of the relatively limited earthquake belts, it must be remembered that most of the earth's surface is aseismic. This has led to the suggestion that the lithosphere is divided into a number of portions or plates, each of which may be moved about over the asthenosphere as a rigid unit, but is only subject to deformation at its junction with another plate. Most geologists now accept that the earth's surface is composed of some seven major lithospheric plates plus a number of smaller ones. Since the boundaries are all marked by earthquake zones, you might like to try dividing the world up into plates on a tracing of Fig. 4.26 and then comparing your answer with a map in a modern textbook such as *Volcanoes*, published by the I.G.S.

The implications of the plate tectonic theory, as it is known, are quite profound and are beyond the scope of the present book. You will, however, find that plate tectonic ideas now underpin much of your other geological reading, in many cases with good justification. In some instances, however, the connections are rather hazy, so you should retain your critical faculties as you read!

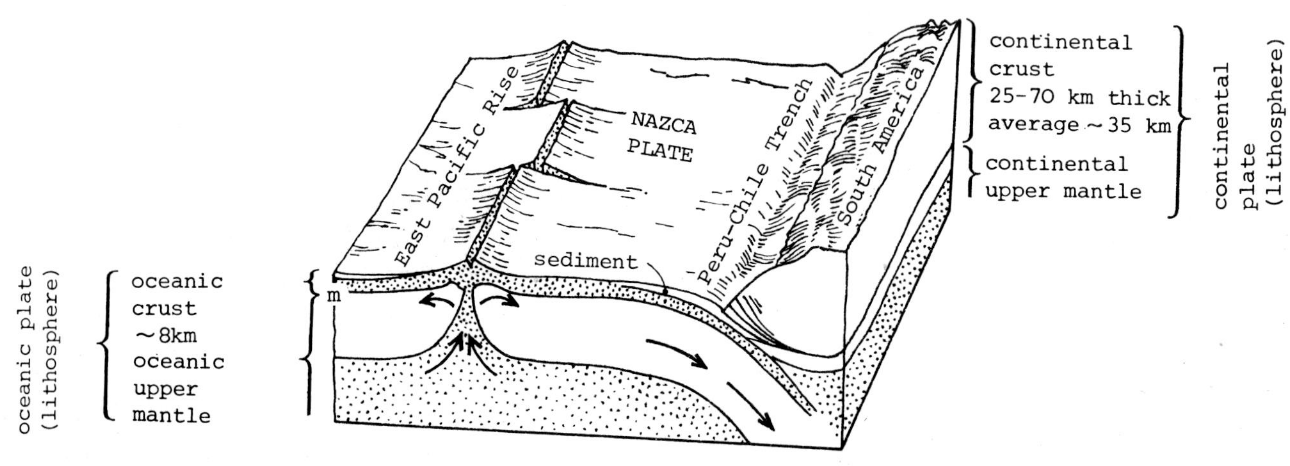

Fig. 4.27 Block diagram of the 'Nazca Plate' in the eastern Pacific.

Answers

<u>Palaeolatitude of Great Britain</u> (Table 3.1 p.14)

Present	$52^{o}N$
Tertiary	$42^{o}N$
Cretaceous	$38^{o}N$
Jurassic	$31^{o}N$
Triassic	$31^{o}N$
Permian	$12^{o}N$
Carboniferous	0°

It appears as though the region stood still between the Triassic and Jurassic, but it is quite probable that there was a change in longitude which cannot be measured from palaeomagnetic specimens.

<u>Seismic refraction travel-time</u> (p. 24)

V 2350 m s^{-1}: Could be shale or sandstone
V_2 5400 m s^{-1}: Could be slate or basic igneous rocks or rock salt

h 67.9 m

<u>Hammer seismic profile</u> (p. 24)

Your analysis of the travel-time curve should have shown:

V_1 = 364 m s^{-1}, V_2 = 1160 m s^{-1}, h = 3.07 m

This could be interpreted in two main ways:

a) 3.07 m of river alluvium overlying Coal Measures siltstones.
b) 3.07 m of dry alluvium above the water table, with saturated alluvium below. Coal Measures siltstones not reached because the line was not long enough.

Note that the velocity of 364 m s^{-1} is close to that of the transmission of sound in air, but it is also known in loosely consolidated gravels.

<u>Seismic wave velocities</u> (p. 28)

1) P wave velocity
2) S waves
3) Velocity is <u>inversely</u> proportional to density.

Further Reading

Dunning, F.W.
Geophysical Exploration, HMSO, 1970. Simple, well illustrated booklet produced by the Geological Museum.

Griffiths, D.H. and King. R.F.
Applied Geophysics for Geologists and Engineers. Pergamon Press, 1981 (2nd edition). Covers all the main methods of geophysical exploration.

Hallam, A.
A Revolution in the Earth Sciences. Oxford University Press, 1973. Historical survey of the growth of current ideas in global geology.

Holmes, A.
Principles of Physical Geology. Revised by D.L. Holmes, Nelson, 1978 (3rd edition).

Khan, M.A.
Global Geology. Wykeham Science Series. Taylor and Francis Group, 1976. General survey of the use of geophysics in global geology.

GEOLOGY TOPICS is made up of six titles focusing on key A level topics:

Palaeoecology explains the interrelationship between living organisms and shows the extent to which fossil assemblages can be used to help understand past environments.
0 582 39972 6

Fossils and Time discusses the various techniques employed to establish a stratigraphy from fossils. Attention is given to the subject of North Sea oil geology.
0 582 39978 5

Geochronology considers stratigraphic principles, followed by an outline of geological methods of establishing an "absolute" time scale.
0 582 39977 7

Geophysics looks at the various methods employed in different branches of geophysics and illustrates the powerful evidence to emerge when the various results are combined. It assesses the importance of this evidence within the context of the search for global resources and the study of plate tectonics.
0 582 39976 9

Igneous Petrology relates the cooling history of magmas to the field of petrographic studies with a basic survey of the physical chemistry involved.
0 582 39975 0

Metamorphism explores the connections between pressure/temperature regimes in the earth's crust and metamorphism. The use of metamorphic textures and structures in determining the history of a rock is an important feature throughout.
0 582 39974 2

Any combination or quantity of titles can be ordered from:

Longman Resources Unit
33-35 Tanner Row
York
YO1 1JP

Longman Oliver & Boyd O&B

Igneous Petrology

P. Kennett & C. A. Ross

First published 1983

Printed in Great Britain by Longman Group Resources Unit, York.